Metal Promoted Cyclocarbonylation Reactions in the Synthesis of Heterocycles

Metal Promoted Cyclocarbonylation Reactions in the Synthesis of Heterocycles

Editor

Laura Antonella Aronica

MDPI • Basel • Beijing • Wuhan • Barcelona • Belgrade • Manchester • Tokyo • Cluj • Tianjin

Editor
Laura Antonella Aronica
Chemistry and Industrial
Chemistry
University of Pisa
Pisa
Italy

Editorial Office
MDPI
St. Alban-Anlage 66
4052 Basel, Switzerland

This is a reprint of articles from the Special Issue published online in the open access journal *Catalysts* (ISSN 2073-4344) (available at: www.mdpi.com/journal/catalysts/special_issues/ MPCRSH_catalsysts).

For citation purposes, cite each article independently as indicated on the article page online and as indicated below:

LastName, A.A.; LastName, B.B.; LastName, C.C. Article Title. *Journal Name* **Year**, *Volume Number*, Page Range.

ISBN 978-3-0365-3774-0 (Hbk)
ISBN 978-3-0365-3773-3 (PDF)

© 2022 by the authors. Articles in this book are Open Access and distributed under the Creative Commons Attribution (CC BY) license, which allows users to download, copy and build upon published articles, as long as the author and publisher are properly credited, which ensures maximum dissemination and a wider impact of our publications.

The book as a whole is distributed by MDPI under the terms and conditions of the Creative Commons license CC BY-NC-ND.

Contents

Laura Antonella Aronica
Metal Promoted Cyclocarbonylation Reactions in the Synthesis of Heterocycles
Reprinted from: *Catalysts* **2022**, *12*, 353, doi:10.3390/catal12040353 1

Vinothkumar Ganesan and Sungho Yoon
Cr-Phthalocyanine Porous Organic Polymer as an Efficient and Selective Catalyst for Mono Carbonylation of Epoxides to Lactones
Reprinted from: *Catalysts* **2020**, *10*, 905, doi:10.3390/catal10080905 5

Pavol Lopatka, Michal Gavenda, Martin Markovič, Peter Koóš and Tibor Gracza
Flow Pd(II)-Catalysed Carbonylative Cyclisation in the Total Synthesis of Jaspine B
Reprinted from: *Catalysts* **2021**, *11*, 1513, doi:10.3390/catal11121513 17

Sijia Chen, Chongguo Jiang, Nan Zheng, Zhen Yang and Lili Shi
Evolution of Pauson-Khand Reaction: Strategic Applications in Total Syntheses of Architecturally Complex Natural Products (2016–2020)
Reprinted from: *Catalysts* **2020**, *10*, 1199, doi:10.3390/catal10101199 33

Xiao-Qiang Hu, Zi-Kui Liu and Wen-Jing Xiao
Radical Carbonylative Synthesis of Heterocycles by Visible Light Photoredox Catalysis
Reprinted from: *Catalysts* **2020**, *10*, 1054, doi:10.3390/catal10091054 59

Gianluigi Albano and Laura Antonella Aronica
From Alkynes to Heterocycles through Metal-Promoted Silylformylation and Silylcarbocyclization Reactions
Reprinted from: *Catalysts* **2020**, *10*, 1012, doi:10.3390/catal10091012 85

 catalysts

Editorial

Metal Promoted Cyclocarbonylation Reactions in the Synthesis of Heterocycles

Laura Antonella Aronica

Dipartimento di Chimica e Chimica Industriale, University of Pisa, Via G. Moruzzi 13, 56124 Pisa, Italy; laura.antonella.aronica@unipi.it

Citation: Aronica, L.A. Metal Promoted Cyclocarbonylation Reactions in the Synthesis of Heterocycles. *Catalysts* 2022, *12*, 353. https://doi.org/10.3390/catal12040353

Received: 18 March 2022
Accepted: 18 March 2022
Published: 22 March 2022

Publisher's Note: MDPI stays neutral with regard to jurisdictional claims in published maps and institutional affiliations.

Copyright: © 2022 by the author. Licensee MDPI, Basel, Switzerland. This article is an open access article distributed under the terms and conditions of the Creative Commons Attribution (CC BY) license (https://creativecommons.org/licenses/by/4.0/).

Oxygen and nitrogen heterocycle systems are found in a vast number of natural substrates and biologically active molecules such as antimycotics, antibiotics, antitumors and antioxidants, in addition to pigments and fluorophores. Therefore, several procedures dedicated to the building of such heterocycles have been developed. Many of them are based on the cyclization of suitable substrates [1–3], multi-component reactions [4,5] and ring expansion processes [6,7]. In this field, metal-catalysed cyclocarbonylative reactions represent atom-economical and efficient methods for the synthesis of several functionalized compounds. Indeed, when the cyclization reaction is performed under CO pressure, the potentiality of the process is enhanced, since the formation of the ring takes place with the contemporary introduction of the carbonyl functional group. An improvement in the field of carbonylation reactions is represented by the substitution of the CO gas by surrogates, molecules which are able to generate the carbon monoxide inside the reaction vessel or may act as CO synthons [8–10].

The present Special Issue collected two research articles and three reviews focused mainly on the preparation of different heterocyclic compounds such as lactones, furans, epoxides, chromanes, chromanones, chromenones, indolinones, tetrahydroquinolines, quinolinones, lactams, benzoimidazoles and benzooxazoles via cyclocarbonylation reactions.

The first research article [11] concerns the selective monocarbonylation of epoxides into the corresponding lactones. The reactions were carried out in the presence of a chromium (III)-phthalocyanine derivative connected to a porous organic polymer. The catalyst was highly effective in promoting the ring expansion reaction and preliminary tests indicated that the catalyst can be reused without losing its catalytic activity.

The preparation of bicyclic lactones has been investigated in the second research article [12] as a key step for the total synthesis of Jaspine B, which has shown promising biological activity as an antitumor against several types of cancer cells. In order to avoid the use of dangerous carbon monoxide gas, the authors have developed a new protocol for Pd-catalysed carbonylation reactions based on the use of iron pentacarbonyl $Fe(CO)_5$ as a CO surrogate. After the optimization of the reaction sequence under batch conditions, the carbonylation reactions were performed also in a flow rector that provided the desired bicyclic lactones in comparable yields to standard batch conditions.

Metal-mediated cyclizations are important transformations in a natural product total synthesis. In the first review [13] of this Special Issue, the Co, Rh, and Pd catalysed Pauson–Khand reactions (PKR) published in the last five years have been summarized. In particular, their application to the synthesis of cyclopentenone and lactone-containing structures have been highlighted. In many examples, the carbonyl moiety has been inserted, employing metal carbonyl compounds as a masked CO source, through the transition metal decarbonylation to in situ generated CO ($Co_2(CO)_8$, $Mo(CO)_3(DMF)_3$, $Rh(CO)_2Cl]_2$). The hetero-Pauson–Khand reaction has also been considered, since it represents a useful tool for the generation of bicyclic γ-butyrolactones and unsaturated lactams. The final part of the review is focused on the synthesis of natural macrocyclic compounds containing cyclopentenone motif.

A different approach to the synthesis of carbonyl-containing oxygen and nitrogen heterocycles based on visible light photocatalytic radical carbonylation has been summarized in the second review [14] of this Special Issue. Acyl radicals serve as the key intermediates in these transformations and can be generated from the addition of alkyl or aryl radicals to carbon monoxide (CO), or from various acyl radical precursors such as aldehydes, carboxylic acids, anhydrides, acyl chlorides or α-keto acids. The discussion of the literature is organized based on the types of acyl radical precursors, and an exhaustive analysis of the transformations leading to the different heterocycles is reported, with particular attention to the mechanistic aspects.

The last review [15] is focused on the synthesis of heterocyclic rings of different sizes, nature and potentialities containing both a silyl and a carbonyl moiety. Intramolecular silylformylation and silylcarbocyclization reactions are the key step for the cyclocarbonylation to occur. The content of this review is divided into two sections: the first is dedicated to a detailed description of intramolecular silylformylation reactions with the corresponding synthesis of oxa- and aza-silacyclane, while the second is centred on the silylcarbocyclization of functionalized acetylenes. In each section, particular emphasis is given to the heterocycles which can be obtained, as well as a special look into used metal catalysts.

In summary, the contribution of the articles collected in this Special Issue will be stimulating for those authors working in the field of heterocycles synthesis and will provide a valuable guide to develop new innovative methodologies for cyclocarbonylative reactions performed under batch, flow and photoredox catalytic conditions, with particular attention to the use of CO surrogates or equivalent synthons.

Finally, I would like to thank all authors for their valuable contributions.

Funding: This research received no external funding.

Conflicts of Interest: The authors declare no conflict of interest.

References

1. Albano, G.; Aronica, L.A. Potentiality and Synthesis of O- and N-Heterocycles: Pd-Catalyzed Cyclocarbonylative Sonogashira Coupling as a Valuable Route to Phthalans, Isochromans, and Isoindolines. *Eur. J. Org. Chem.* **2017**, *2017*, 7204–7221. [CrossRef]
2. Aronica, L.A.; Albano, G. Supported Metal Catalysts for the Synthesis of N-Heterocycles. *Catalysts* **2022**, *12*, 68. [CrossRef]
3. Sartori, S.K.; Diaz, M.A.N.; Diaz-Munoz, G. Lactones: Classification, synthesis, biological activities, and industrial Applications. *Tetrahedron* **2021**, *84*, 132001. [CrossRef]
4. Borah, B.; Dwivedi, K.D.; Chowhan, R. 4-Hydroxycoumarin: A Versatile Substrate for Transition-metal-free Multicomponent Synthesis of Bioactive Heterocycles. *Asian J. Org. Chem.* **2021**, *10*, 3101–3126. [CrossRef]
5. Qiao, S.; Zhang, N.; Wu, H.; Hanas, M. Based on MFe_2O_4 NPs catalyzed multicomponent reactions: Green and efficient strategy in synthesis of heterocycles. *Synth. Commun.* **2021**, *51*, 2873–2891. [CrossRef]
6. Singh, G.S.; Sudheesh, S.; Keroletswe, N. Recent applications of aziridine ring expansion reactions in heterocyclic synthesis. *Arkivoc* **2018**, *1*, 50–113. [CrossRef]
7. Malapit, C.A.; Howell, A.R. Recent Applications of Oxetanes in the Synthesis of Heterocyclic Compounds. *J. Org. Chem.* **2015**, *80*, 8489–8495. [CrossRef] [PubMed]
8. Khedkar, M.V.; Khan, S.R.; Lambat, T.L.; Chaudhary, R.G.; Abdala, A.A. CO Surrogates: A Green Alternative in Palladium-Catalyzed CO Gas Free Carbonylation Reactions. *Curr. Org. Chem.* **2020**, *24*, 2588–2600. [CrossRef]
9. Chen, Z.; Wang, L.C.; Wu, X.F. Carbonylative synthesis of heterocycles involving diverse CO surrogates. *Chem. Commun.* **2020**, *56*, 6016–6030. [CrossRef] [PubMed]
10. Panda, B.; Albano, G. DMF as CO Surrogate in Carbonylation Reactions: Principles and Application to the Synthesis of Heterocycles. *Catalysts* **2021**, *11*, 1531. [CrossRef]
11. Ganesan, V.; Yoon, S. Cr-Phthalocyanine Porous Organic Polymer as an Efficient and Selective Catalyst for Mono Carbonylation of Epoxides to Lactones. *Catalysts* **2020**, *10*, 905. [CrossRef]
12. Lopatka, P.; Gavenda, M.; Markovic, M.; Koóš, P.; Gracza, T. Flow Pd(II)-Catalysed Carbonylative Cyclisation in the Total Synthesis of Jaspine B. *Catalysts* **2021**, *11*, 1513. [CrossRef]
13. Chen, S.; Jiang, C.; Zheng, N.; Yang, Z.; Shi, L. Evolution of Pauson-Khand Reaction: Strategic Applications in Total Syntheses of Architecturally Complex Natural Products. *Catalysts* **2020**, *10*, 1199. [CrossRef]

14. Hu, X.-Q.; Liu, Z.-K.; Xiao, W.-J. Radical Carbonylative Synthesis of Heterocycles by Visible Light Photoredox Catalysis. *Catalysts* **2020**, *10*, 1054. [CrossRef]
15. Albano, G.; Aronica, L.A. From Alkynes to Heterocycles through Metal-Promoted Silylformylation and Silylcarbocyclization Reactions. *Catalysts* **2020**, *10*, 1012. [CrossRef]

Article

Cr-Phthalocyanine Porous Organic Polymer as an Efficient and Selective Catalyst for Mono Carbonylation of Epoxides to Lactones

Vinothkumar Ganesan and Sungho Yoon *

Department of Chemistry, Chung-Ang University, 84, Heukseok-ro, Dongjak-gu, Seoul 06974, Korea; vinothcau@cau.ac.kr
* Correspondence: sunghoyoon@cau.ac.kr

Received: 19 July 2020; Accepted: 6 August 2020; Published: 8 August 2020

Abstract: A facile, one-pot design strategy to construct chromium(III)-phthalocyanine chlorides (Pc'CrCl) to form porous organic polymer (POP-Pc'CrCl) using solvent knitting Friedel-Crafts reaction (FCR) is described. The generated highly porous POP-Pc'CrCl is functionalized by post-synthetic exchange reaction with nucleophilic cobaltate ions to provide an heterogenized carbonylation catalyst (POP-Pc'CrCo(CO)$_4$) with Lewis acid-base type bimetallic units. The produced porous polymeric catalyst is identical to that homogeneous counterpart in structure and coordination environments. The catalyst is very selective and effective for mono carbonylation of epoxide into corresponding lactone and the activities are comparable to those observed for a homogeneous Pc'CrCo(CO)$_4$ catalyst. The (POP-Pc'CrCo(CO)$_4$) also displayed a good catalytic activities and recyclability upon successive recycles.

Keywords: Cr-phthalocyanine; porous organic polymer; Friedel–Crafts reaction; heterogeneous catalysis; carbonylation; β-lactones; catalyst recyclability

1. Introduction

β-Lactones are an important class of energetically favored four-membered heterocycles with prevalent utilities in the chemical industry, since they are crucial intermediates for the production of various derivatives of β-hydroxy acids, biodegradable poly(β-hydroxyalkanotes), succinic anhydrides, and acrylic acids [1–9]. Their inherent ring strain facilitates excellent reactivity, allowing them to undergo a range of transformations to provide products with a variety of applications ranging from polymer chemistry to natural product synthesis [4]. However, synthetic routes to β-lactones are limited. Recently, ring-expansion epoxide carbonylation utilizing inexpensive C1 sources have emerged as a convenient and direct method to produce β-lactones with good atom economy [10].

Although there have been numerous reports of ring-expansion carbonylation catalysts, well-defined Lewis acid–base ion pairing catalysts of the common type [Lewis acid]$^+$ [Co(CO)$_4$]$^-$ have demonstrated high efficiency for these transformations [11–13]. Among the reported Lewis acid–base pair catalysts, the porphyrin-based [OEPCr(THF)$_2$]$^+$ [Co(CO)$_4$]$^-$ (OEP = Octaethylporphyrinato, THF = tetrahydrofuran) catalyst has demonstrated high reactivity and high selectivity toward mono carbonylation under homogeneous conditions [14]. However, tedious catalyst synthesis and product separation have limited the use of this catalytic system and motivated the search for viable alternatives, including heterogeneous systems [15–17]. In addition to the heterogenization of catalysts to improve recyclability, facile synthesis of such catalytic systems is being actively researched [18–20].

Phthalocyanine (Pc), which is a porphyrinoid analogue, is easily synthesized with excellent yields and could serve as an alternative for porphyrin systems; because of their planar tetradentate dianionic ligation, phthalocyanines are excellent structural analogs to porphyrins and are synthetically facile. Recently, we demonstrated that a catalyst generated in situ from commercial (AlPcCl) and $Co_2(CO)_8$ displayed excellent activity for mono and double carbonylation [21]. But, the selectivity toward β-lactones was very poor, hence, Lewis acidic Al^{3+} containing [Lewis acid]$^+$ [Co(CO)$_4$]$^-$ type ion pairing catalysts are proven to be active also for double carbonylation and generally resulting in a mixture of β-lactones and anhydrides [3,21,22]. Therefore, Cr^{3+} containing [PcCr(III)]$^+$ [Co(CO)$_4$]$^-$ type catalyst could be a suitable Lewis acidic part for selective monocarbonylation of epoxides into β-lactones [12,13,23]. However, partial solubility of Pc metal complex due to intermolecular π–π stacking interactions leads to reduced collision between the Pc metal complex and substrate (epoxide) resulting in low catalytic activity. These enforced further structural tunings on the Pc ring to improve the catalyst solubility and activity by controlling such a π–π stacking interactions; in addition, the intensely colored Pc metal complexes were difficult to separate from the reaction mixture [21,24,25].

Immobilization of soluble Pc metal complexes on a flexible POP addresses both, solubility and separation issues, by providing a heterogeneous catalytic system. In this regard, we strategically designed and synthesized a new phthalocyanine chromium(III) chloride complex (Pc'Cr(III)Cl) and heterogenized using a simple, one-pot solvent-knitting Friedel–Crafts reaction (FCR); the resulting complex was functionalized with [Co(CO)$_4$]$^-$ anion to generate a highly active and recyclable [POP-Pc'Cr(III)]$^+$ [Co(CO)$_4$]$^-$ catalytic system.

2. Results and Discussion

2.1. Synthesis and Characterization of POP-Pc'Cr(III)Cl

A synthetic strategy of POP-Pc'Cr(III)Co(CO)$_4$ (**4**) is shown in Scheme 1. At first, the ligand **1** is synthesized by the mild base-catalyzed condensation [26]. The monomer Pc'Cr(III)Cl (**2**) was synthesized with excellent yield according to the modified literature procedure and characterized by FTIR and UV-Visible spectroscopic techniques, further confirmed by high-resolution mass spectrometry as shown in Figures S1–S4 [27,28]. The substituted 2-isopropylphenolic group not only improves the solubility of the monomer **2** but also acts as a knitting group through covalent linkages by AlCl$_3$-catalyzed FCR using methylene dichloride both as a crosslinker and as a solvent. The resulting dark green color porous organic polymer POP-Pc'CrCl (**3**) is stable under open atmosphere conditions and not soluble in most commonly used organic solvents as a result of extensive cross-linking [18,19]. The compositional homogeneity and the surface topography of POP **3** was probed by a scanning electron microscope (SEM) and a transmission electron microscope (TEM) analysis. Figure 1a,b shows the SEM and TEM images of **3** respectively, as an aggregate of polydisperse spherical shape particles of 1 μm size average (Figures S5 and S6). Energy dispersive X-ray (EDS) analysis shows the relative abundance of constituent elements throughout the Pc' polymeric matrix indicating uniform distribution of elements after polymerization (Figure S5) [29]. A powder X-ray diffraction analysis of the resulting polymer showed that a wide peak at 2θ = 13.1° attributes to the construction of amorphous polymeric material (Figure S7a). Subsequently, the absence of distinctive sharp monomeric diffraction peaks (Figure S7b) suggests that the solvent knitting polymerization is thoroughly completed and the resulting POP is free from crystalline monomer residues [18]. The thermal endurance of the resulting polymeric material was characterized by thermogravimetric analysis (TGA) as shown in Figure S8, the polymeric material was stable up to 400 °C indicating possible outstanding thermal stability under harsh reaction conditions.

The porosity of POP-Pc'CrCl, **3** was investigated by N$_2$ sorption measurement carried out at 77 K. The Pc'-based polymer **3** exhibited characteristic type-I adsorption isotherms (as per IUPAC adsorption isotherms classification), as depicted in Figure 1c [30]. A steep N$_2$ uptake at a lower relative pressure (P/P$_0$ = 0–0.1) region is attributed to the microporous character of polymer **3**, whereas the hysteresis loop behavior throughout the range of relative pressure can be attributed to the existence of mesoporosity. The BET (Brunauer–Emmett–Teller) surface area of the polymeric material is 725 m^2 g^{-1}

and the total pore volume is 0.388 cm^3 g^{-1} (Figure 1c and Figure S9a). These porosity results are comparable with other reported Pc-based porous polymers and indicate high surface area, which enables larger exposure of active sites per unit mass of the material; they also confirm the distribution of high grade porous structure that can accommodate bulky functional groups via the labile chloro ligand for specific catalytic applications [18,31–33].

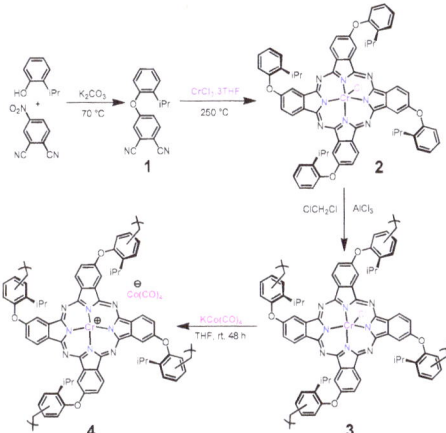

Scheme 1. Synthesis of catalyst **4** by the Friedel-Crafts reaction (FCR).

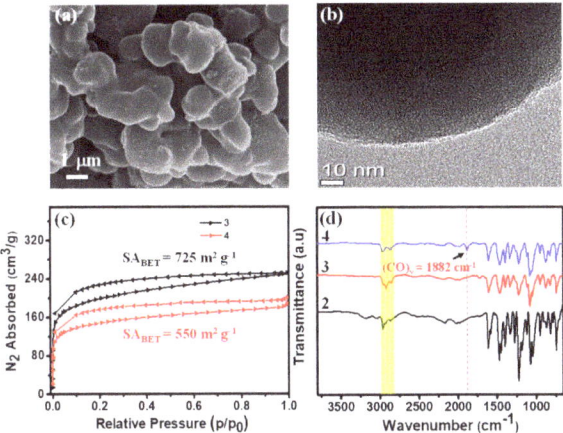

Figure 1. (a) SEM image of POP-Pc'CrCl, (b) TEM image of **3**, (c) N$_2$ sorption isotherms of **3** and **4** at 77 K, and (d) FTIR spectra of **2**, **3**, and **4**.

The chemical structure of the resulting polymer was analyzed using FTIR spectroscopy. Comparative FTIR spectral analysis of the monomeric complex and the formed porous polymer was performed to evaluate the structural integrity of the resulting polymers, as displayed in Figure 1d. IR peaks appearing in the range of 3068–2856 cm^{-1} are attributed to –C=C–H stretching vibrations of the aromatic functional groups of the phthalocyanine complex and polymer **3**, as well as to the newly formed methylene crosslinking bridges [34]. The peaks at 1610, 1470, and 1390 cm^{-1} can be assigned for C=C, C–N, and C=N stretching vibrations, respectively, of the Pc' ring, which contains benzene, aza, and pyrrole functional groups; these are present in the formed polymers as well as the parent monomers [31,35,36]. These FTIR spectroscopic analysis shows that the resulting POP retains most of the feature peaks of its corresponding monomer and is consistent with the anticipated polymeric structure. Thus, these analysis results unambiguously verify the direct heterogenization of Pc'Cr(III)Cl complex by the one-pot FCR to produce very stable, and heterogeneous porous polymeric materials.

2.2. Synthesis and Characterization of POP-Pc'Cr(III)Co(CO)$_4$

Heterogeneous phthalocyanine polymer matrix is a probable candidate for Lewis-acid-enabled catalytic transformations. Particularly, the Pc'Cr(III)Cl complex resembles TPPCr(III)Cl (TPP = Tetraphenylporphyrinato), the best candidate for the carbonylation of epoxides with an additional Lewis base incorporation [13,37]. Interestingly, polymeric material 3 can be incorporated with an appropriate base via its labile Cl$^-$ anions in order to form a Lewis acid–base ion pair catalyst [12–14]. Accordingly a metathesis reaction of Co(CO)$_4$$^-$ anions can replace the labile Cl$^-$ ions to generate a heterogeneous bimetallic frustrated Lewis acid-base ion pair type catalyst ([Lewis acid]$^+$[Co(CO)$_4$]$^-$)to promote the epoxide ring-expansion carbonylation [17–20]. As such, polymer 3 was treated with excess KCo(CO)$_4$ to generate the heterogeneous epoxide carbonylation catalyst [POP-Pc'Cr]$^+$ [Co(CO)$_4$]$^-$ (4) [12–14]. At first the resulting catalyst 4 was characterized by FTIR spectroscopic technique. Compared to polymer 3 (contains Cl$^-$), the Co(CO)$_4$$^-$ anions exchanged catalyst 4 exhibits a strong new absorption peak at 1882 cm^{-1} (Figure 1d). This peak is characteristic of typical ν(CO) from newly exchanged tetrahedral Co(CO)$_4$$^-$ ions, consistent with that of previously reported well-defined homogeneous Cr-containing [Lewis acid]$^+$ [Co(CO)$_4$]$^-$-type catalysts [17,19]. SEM and TEM images of the catalyst indicate no morphology change after Co(CO)$_4$$^-$ anion exchange. Subsequently, EDS analysis confirms the incorporation of Co into the polymeric frameworks along with other constituent elements, distributed uniformly all over the polymer(Figure 2a,b, Figures S10 and S11). Atomic absorption spectroscopy (AAS) and inductively coupled plasma–optical emission spectroscopy (ICP-OES) revealed that the Co and Cr contents were 1.78 and 3.63 wt%, respectively, against, 4.63 and 4.09 wt% calculated for Co and Cr, respectively, in well-defined homogeneous catalyst. The molar ratio of the Cr/Co content in catalyst 4 is 1.8 (determined by ICP-AAS), indicating partial exchange of Cl$^-$ ion and a part of Lewis-acidic Cr^{3+} remains combined with Cl$^-$ ions; they could be buried inside the microporous channels and/or inaccessible for cobaltate exchange. Limited molecular exchange of cobaltate ion pairs is consistent with SEM-EDS analysis and also reported previously [17,19].

The coordination environment of the catalyst 4 metal species was characterized using X-ray photoelectron spectroscopy (XPS). The XPS peak for Cr 2p shows a characteristic doublet at 577.21 and 586.70 eV as shown in Figure 2c, which matches well with the structural analogues, the POP-TPP-supported Cr(III) species, and analogous metal center on porous organic networks [17,19,20,38]. As shown in Figure 2d, the XPS peaks for Co species are detected at 796.90 eV and 781.55 eV along with the typical shoulder is for the Co 2p$_{1/2}$ and Co 2p$_{3/2}$ orbitals of the Co(CO)$_4$$^-$ species, respectively. The observed Co XPS peaks values are also consistent with those of Co(CO)$_4$$^-$-exchanged similar TPPAl, CTF-Al(OTf), and TPPCr heterogeneous catalysts [16–20]. Finally, the porosity retention is evident from TEM images (Figure S11) and N$_2$ gas sorption measurements carried out at 77 K, which afford type-I isotherms and exhibiting hysteresis loop behavior, displaying a combination of micro and mesoporosity (Figure 1c). However, the BET surface area is reduced to 550 m^2 g^{-1} for catalyst 4, and a decreased total pore volume of 0.28 cm^3 g^{-1} (related to the parent polymeric network) is observed (Figure 1c and Figure S9b). This suggests that the exchanged Co(CO)$_4$$^-$ anions partly occupy the porous channels of polymeric network, thereby decreasing the total available pore volume as observed previously [17–19]. Nevertheless, the catalyst maintains a porous structure to allow the substrate epoxide and the product β-butyrolactone molecules to diffuse over the Lewis acid–base-ions paired porous channels.

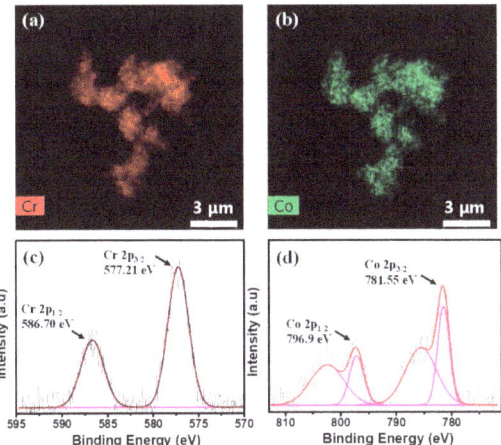

Figure 2. STEM-EDS mapping image of (**a**) Cr atoms and (**b**) Co atoms and X-ray photoelectron profiles of **4** for deconvoluted (**c**) Cr2p and (**d**) Cobalt 2p core level.

2.3. Carbonylation Activity of POP-Pc'Cr(III)Co(CO)$_4$

Catalyst **4** was tested for carbonylation catalytic activity in a 50 mL stainless steel custom-made one inch tubular reactor. Propylene oxide (PO) was used as an epoxide substrate with CO under 6 MPa pressure. Various solvents were tested, since carbonylation is affected by the type of solvent [14,22]. The crude reaction mass was analyzed by ^{1}H NMR spectral analysis using an internal standard naphthalene; the results are summarized in Table 1. Among the solvents screened, weakly coordinating DME is the most active solvent system for PO carbonylation to β-butyrolactone with >99% conversion and selectivity (entries 1–4), consistent with earlier reports for POP-TPPCrCo(CO)$_4$ and analogs [17,19,20]. Using similar reaction conditions, we evaluated the catalytic activity of homogenous well-defined [Pc'Cr][Co(CO)$_4$]; we observed conversion of >99% and selectivity of 99% toward β-butyrolactone (entry 5). The activity was tested with a higher PO ratio, and the yield was reduced to a ratio of 200 (entry 6). Reactions were performed for 12 h and 1 h under same reaction conditions (entry 7 and 8, respectively) to get initial rates of conversion, and the carbonylation yields of 70 and 22% were observed, respectively, with a site time yield of 44 h^{-1}. Finally, the activity was tested at room temperature; 40% yield was achieved with a substrate/catalyst ratio of 100 (entry 9). The activity of POP-Pc'Cr(III)Cl was also evaluated under the same reaction conditions; we observed only PO to polyether conversion, owing to the absence of a Lewis base for carbonyl group insertion (entry 10) [21].

Table 1. Carbonylation activity of catalyst 4.

Entry [a]	Catalyst	Solvent	Epoxide/Co Ratio [b]	T (°C)	Time (h)	Yield [c] (%)	Lactone (%)	Acetone (%)
1	4	DME	200	60	24	>99	>99	<1
2	4	THF	200	60	24	50	99	1
3	4	1,4-dioxane	200	60	24	52	>99	<1
4	4	Toluene	200	60	24	75	>99	<1
5	[Pc'Cr][Co(CO)$_4$]	DME	200	60	24	>99	99	1
6	4	DME	400	60	24	52	>99	<1
7	4	DME	200	60	12	70	>99	<1
8	4	DME	200	60	1	22	>99	<1
9	4	DME	100	30	24	40	>99	trace
10	3	DME	200	60	24	12 [d]		

[a] Reactions performed in DME solution of epoxide (1.8 M) under 6 MPa CO pressure at respective temperature. The mixture was stirred in a preheated oil bath to maintain respective temperature. [b] Calculated based on ICP-AAS value for Co content. [c] Determined by ^{1}H-NMR spectra with an internal standard naphthalene. [d] Polyether was formed.

Before testing recyclability, the heterogeneous nature of catalyst **4** was examined using a hot filtration test; a suspension of catalyst **4** in DME solvent was stirred at 60 °C for 6 h, and the treated catalyst was separated by filtration [15,18]. The dried solid catalyst and filtrate were subjected to the standard carbonylation conditions (2 mol% of catalyst, 6 MPa of CO, 60 °C, and 24 h reaction time) separately. Only the separated solids promoted epoxide carbonylation: no significant epoxide conversion was observed in the presence of the filtrate under the same conditions. This confirms that catalyst **4** retains heterogeneity [18,19]. Finally, catalyst **4** was evaluated for recyclability. Epoxide carbonylation was carried out with catalyst **4** at 30 °C temperature for 24 h under 6 MPa CO pressure in DME solvent. After the reaction, the reaction mixture was filtered inside a glove box to isolate the solid catalyst, which was then washed with dry DME and dried under vacuum; the dried catalyst was used for successive cycles. ^{1}H NMR spectral analysis of the recovered filtrate was conducted to evaluate the recycling ability of the catalyst, as listed in Table 2. The activity was reduced from complete conversion to 98% in the second cycle. The activity decreased further to 85 ± 6% in the third cycle. After the third cycle, the catalyst was analyzed by SEM-EDS to understand the reason for the decreased activity. SEM-EDS analysis of catalyst **4** after third cycle shows no changes in the catalyst morphology, but did reveal an increase in the Cr/Co ratio due to reduced Co content in the catalyst. As shown in Figure S12, the ratio of Cr/Co increased from 1.8:1 to 3.5:1 after the third cycle [16–19]. This suggests that the decreased Co content in the catalyst causes reduced activity during recycling [18,19]. Notably, the spent catalyst (after three cycles) was subjected to treatment with KCo(CO)$_4$ for regeneration [17–19]. The regenerated catalyst revealed restoration of the catalytic activity upon testing. This further substantiates that leaching of Co causes catalyst deactivation during recycling and the catalyst activity can be restored by treating with cobaltate ions to replenish the activity. Thus, POP-Pc'CrCo(CO)$_4$, prepared via the solvent-knitting FCR, is an efficient and recyclable heterogeneous catalyst.

Table 2. Recyclability of 4.

Cycle	Yield (%)	Selectivity (%) β-Butyrolactone/Acetone
1	>99	>99/<1
2	98	>99/<1
3	85 ± 6	>99/<1
4 *	98	>99/<1

Reaction conditions: catalyst 2 mol%, 6 MPa of CO pressure, 30 °C, DME solvent. The PO conversion was determined by ^{1}H NMR spectra measured with internal standard naphthalene. * regenerated catalyst.

3. Experimental Section

3.1. Materials and Methods

All chemicals and reagents were procured from commercial dealers and used as received unless otherwise mentioned. Chemicals 4-nitrophthalonitrile, 2-isopropylphenol, anhydrous aluminum chloride ($AlCl_3$), dicobaltoctacarbonyl ($Co_2(CO)_8$), tetrahydrofuran (THF), dimethanoxyethane (DME), 1,4-dioxane, toluene, and propylene oxide (PO) were purchased from Sigma-Aldrich (Seoul, Korea). The solvent THF, DME, 1,4-dioxane, and toluene were distilled over sodium/benzoquinone and PO was distilled over CaH_2 under argon atmosphere. Deuterated solvents were purchased from Cambridge Isotope Laboratories, Inc. (T&J Tech Inc, Seoul, Korea). Research grade carbon monoxide was purchased from Air Liquide Korea Co., Ltd. (Seoul, Korea) with 99.998% purity and used as received. The $KCo(CO)_4$ was synthesized according to the reported procedure [39,40]. All manipulations of air and moisture sensitive compounds were carried out inside the glove box under argon atmosphere. Attenuated total reflectance infrared (ATR-IR) measurements were carried out on a Nicolet iS 50 (Thermo Fisher Scientific, Waltham, MA, USA). Scanning electron microscopy (SEM) and energy-dispersive spectroscopy (EDS) measurements were performed using a JEM-7610F (JEOL Ltd., Tokyo, Japan) operated at an accelerating voltage of 20.0 kV. The morphology of the prepared catalysts was observed by a transmission electron microscope Tecnai G2 (FEI Company, Hillsboro, OR, USA). TEM-EDX elemental mapping was obtained with transmission emission microscopy Talos F200X (Thermo Fisher Scientific, Waltham, MA, USA). The X-ray photoelectron spectrum (XPS) was obtained using K-Alpha X-ray photoelectron spectrometer (Thermo Fisher Scientific, Waltham, MA, USA). The binding energies were corrected by the C1s peak from carbon contamination to 284.6 eV. The metal content of the catalysts was analyzed by inductively coupled plasma optical emission spectroscopy (ICP-OES) (iCAP 6000 series, Thermo Fisher Scientific, Waltham, MA, USA) using a microwave-assisted acid digestion system (MARS6, CEM/USA). Samples (~20.0 mg) were digested in a mixture of conc. HCl (20.0 mL) and conc. H_2SO_4 (10.0 mL) solution under microwave rays at 210 °C for 60 min (ramp rate = 7 °C/min). N_2 adsorption-desorption measurements were conducted in an automated gas sorption system (Belsorp II mini, MicrotracBEL, Osaka, Japan) at 77 K; the samples were degassed for 12 h at 80 °C before the measurements. The Brunauer-Emmett-Teller (BET) and Barrett-Joyner-Halenda (BJH) methods were used for calculating the surface areas and pore size distributions, respectively. Powder X-ray diffraction (PXRD) was measured on a RIGAKU D/Max 2500 V using CuKα radiation. ^{1}H and ^{13}C NMR were measured on a 600 MHz Varian VNS NMR spectrometer (Varian, Inc., CA, USA) and 400 MHz NMR spectrometer Bruker Avance III 400 (Bruker Korea Co., Ltd., Seoul, Korea). Simultaneous DSC-TGA instrument (TA instruments, New Castle, DE, USA) was used for the thermogravimetric analysis (TGA) with a heating rate of 10 °C/min from 25 °C to 800 °C under nitrogen atmosphere. UHR-MS measurements were performed on Bruker compact mass spectrometer (Bruker Korea Co., Ltd., Seoul, Korea).

3.2. Synthesis of Pc′ Ligand

4-Nitrophthalonitrile (5.01 g, 0.03 mol), 2-isopropylphenol (4.34 g, 0.03 mol), and K_2CO_3 (6.00 g, 0.04 mol) were stirred in anhydrous N,N-dimethylformamide (20 mL) at 52 °C for 24 h under N_2 atmosphere. A dark brown solution was obtained and was poured into ice-cold water (200 mL). The resulting brown precipitate was filtered, washed with water, and dissolved in dichloromethane (200 mL); the organic phase was purified by water extraction (3 × 100 mL). The desired product was purified by flash column chromatography (silica gel; hexane/ethyl acetate: 10:1) and recrystallized in hot methanol to obtain a pale white crystalline solid in 90% yield. FTIR: (cm^{-1}) 3085, 2977, 2870, 2233, 1590, 1481, 1446, 1415. 1311, 1280, 1246, 1218, 1184, 1084, 952, 872, 852, 775, 752; ^{1}H NMR (600 MHz, CDCl$_3$, ppm) δ 7.71 (d, J = 8.7 Hz, 1H), 7.42 (dd, J = 7.5, 1.8 Hz, 1H), 7.32–7.25 (m, 2H), 7.23 (d, J = 2.5 Hz, 1H), 7.18 (dd, J = 8.7, 2.6 Hz, 1H), 6.93 (dd, J = 7.8, 1.3 Hz, 1H), 3.03 (dt, J = 13.8, 6.9 Hz, 1H), 1.18 (d, J = 6.9 Hz, 6H).^{13}C NMR (151 MHz, CDCl$_3$) δ 162.33 (s), 150.63 (s), 140.91 (s), 135.56 (s), 128.14 (s), 127.90 (s), 127.05 (s), 121.08 (s), 121.03 (s), 120.85 (s), 117.84 (s), 115.55 (s), 115.13 (s), 108.65 (s), 27.34 (s), 23.15 (s).

3.3. Synthesis of Pc′Cr(III)Cl

In a glove box, CrCl$_3$·3THF (0.36 g, 0.96 mmol) and 4-(2-isopropylphenoxy)phthalonitrile (1.00 g, 3.82 mmol) were added to a 20 mL ampoule, which was then sealed under high vacuum. The ampoule was heated at a rate of 60 °C per hour to 250 °C and maintained at the same temperature for 5 h. The ampoule was then cooled to room temperature to obtain a dark product. The product was subsequently removed from the ampoule and purified by Soxhlet extraction using dichloromethane for 48 h to obtain a very dark green crystalline product in 80% yield. FTIR: (cm^{-1}) 3174, 2962, 2865, 1612, 1470, 1396, 1334, 1276, 1222, 1072, 1045, 952, 872, 818, 790, 748; UV-Vis: (THF) λ_{max} 281, 366, 491, 622, 690 nm; HRMS (ESI Q-TOF) m/z calculated $[C_{68}H_{56}CrN_8O_4]^+$ 1100.3830, found $[M-Cl]^+$ 1100.3832.

3.4. Synthesis of POP-Pc′Cr(III)Cl

Under Ar atmosphere, Pc′Cr(III)Cl (1.00 g, 0.87 mmol) was suspended in 40 mL dichloromethane, the reaction mixture was cooled to 0 °C, and fresh anhydrous AlCl$_3$ (1.87 g, 14.07 mmol) was added. The reaction mixture was then stirred at 0 °C for 4 h, 30 °C for 8 h, 40 °C for 12 h, 60 °C for 12 h, and 80 °C for 24 h to obtain a dark-colored polymerized solid suspension. The resulting solid suspension was quenched using 50 mL of a HCl-H$_2$O mixture (v/v = 2:1), washed with water thrice and with ethanol twice, then with THF, methanol, water, acetone, pentane, and ether (100 mL each). It was further purified by Soxhlet extraction with 1:1 methanol/THF for 48 h, and then dried in a vacuum oven at 80 °C for 24 h to obtain a dark green solid. FTIR: (cm^{-1}) 2931, 2854, 1608, 1465, 1392, 1334, 1226, 1080, 1049, 879, 825, 748.

3.5. Synthesis of [POP-Pc′Cr(III)][Co(CO)$_4$]

Inside the glove box, the heterogenized POP-Pc′Cr(III)Cl (1.00 g) was suspended in 10 mL dry THF and was added to a THF solution of KCo(CO)$_4$ (1.02 g). The solution was stirred at room temperature for 48 h, following which the reaction mixture was filtered to remove the dark precipitate, which was washed with THF (3 × 50 mL) and dried under high vacuum for 8 h to yield a dark green solid. FTIR: (cm^{-1}) 2965, 2870, 1882, 1608, 1458, 1396, 1334, 1218, 1080, 1053, 1049, 879, 825, 748.

3.6. PO Carbonylation Procedure

A stainless steel carbonylation reactor was dried overnight and placed inside the glove box. The reactor was charged with the POP-Pc′Cr(III)Co(CO)$_4$ catalyst (0.01 g, 12.16 µmol) and a dimethoxyethane solution of propylene oxide (1.8 M in 2.5 mL, PO/catalyst ratio = 200). The reactor was tightened completely and pressurized to 6 MPa of CO after the removal from the glove box and then placed in a preheated oil bath at 60 °C for 24 h. At 60 °C, the pressure was 6.2 MPa; after completion of the reaction, the reactor was brought to room temperature (pressure ~6 MPa) and cooled in an ice bath, following which CO gas was vented slowly inside the fume hood. The filtrate of the reaction mixture was analyzed by ^{1}H NMR spectroscopy using internal standard naphthalene (*Caution*: carbon monoxide (CO) is a highly toxic gas, should be handled with extreme care inside the well-ventilated hood with a proper CO detector).

4. Conclusions

A new design strategy was presented for the facile synthesis of a chromium(III)phthalocyanine-based porous organic polymer (POP-Pc'CrCl) through a solvent-knitting Friedel–Crafts reaction. The constructed POP-Pc'CrCl has a high porosity with a BET specific surface area of 725 m^2 g^{-1}. When functionalized with cobaltate ([Co(CO)$_4$]$^-$) anions, the resulting heterogenized bimetallic Lewis acid–base ion pair catalyst exhibits epoxide ring-expansion carbonylation activity comparable to that of its homogeneous counterpart with slightly reduced activity during successive recycles which can be replenished upon catalyst regeneration. This new design strategy is useful for the synthesis of soluble metallophthalocyanines and one step construction of porous organic polymer for specific catalytic applications.

Supplementary Materials: The following are available online at http://www.mdpi.com/2073-4344/10/8/905/s1. Figure S1: ^{1}H NMR spectrum of ligand **1** measured in CDCl$_3$. Figure S2: ^{13}C NMR spectrum of ligand **1** measured in CDCl$_3$. Figure S3: UV-Visible spectrum of Pc'Cr(III)Cl (**2**). Figure S4: HR-MS of Pc'Cr(III)Cl (**2**). Figure S5: SEM and EDS mapping images of **3**. Figure S6: TEM images of **3**. Figure S7: Powder X-ray diffraction pattern of **3**. Figure S8: TGA plots of **2** and **3**. Figure S9: BJH pore size distribution graph of **3** and **4**.. Figure S10: SEM and EDS mapping images of **4**. Figure S11: TEM and EDS mapping images of **4**. Figure S12: SEM-EDS images of catalyst **4** after cycle three.

Author Contributions: V.G. and S.Y. designed the experiments. V.G. conducted the experiments. V.G. and S.Y. wrote the original draft. Review and edited by S.Y. Supervision, project administration, and funding acquisition by S.Y. All authors have read and agreed to the published version of the manuscript.

Funding: This work was supported by the C1 Gas Refinery Program (No. 2018M3D3A1A01018006) and ERC program (No. 2020R1A5A1018052) through the National Research Foundation of Korea (NRF) grants funded by the Ministry of Science, ICT and Future Planning, Republic of Korea.

Acknowledgments: We acknowledge the financial support by the C1 Gas Refinery Program (No. 2018M3D3A1A01018006) and ERC program (No. 2020R1A5A1018052) through the National Research Foundation of Korea (NRF) grants funded by the Ministry of Science, ICT and Future Planning, Republic of Korea.

Conflicts of Interest: The authors declare no conflict of interest.

References

1. Dunn, E.W.; Lamb, J.R.; Lapointe, A.M.; Coates, G.W. Carbonylation of Ethylene Oxide to β-Propiolactone: A Facile Route to Poly(3-hydroxypropionate) and Acrylic Acid. *ACS Catal.* **2016**, *6*, 8219–8223. [CrossRef]
2. Rajendiran, S.; Park, G.; Yoon, S. Direct Conversion of Propylene Oxide to 3-Hydroxy Butyric Acid Using a Cobalt Carbonyl Ionic Liquid Catalyst. *Catalysts* **2017**, *7*, 228. [CrossRef]
3. Getzler, Y.D.Y.L.; Kundnani, V.; Lobkovsky, E.B.; Coates, G.W. Catalytic Carbonylation of β-Lactones to Succinic Anhydrides. *J. Am. Chem. Soc.* **2004**, *126*, 6842–6843. [CrossRef] [PubMed]
4. Robinson, S.L.; Christenson, J.K.; Wackett, L.P. Biosynthesis and chemical diversity of β-lactone natural products. *Nat. Prod. Rep.* **2019**, *36*, 458–475. [CrossRef]
5. Wang, Y.C.; Tennyson, R.L.; Romo, D. β-lactones: Intermediates for Natural Product Total Synthesis and New Transformations. *Heterocycles* **2004**, *64*, 605–658. [CrossRef]
6. Li, Z.; Yang, J.; Loh, X.J. Polyhydroxyalkanoates: Opening doors for a sustainable future. *NPG Asia Mater.* **2016**, *8*, e265. [CrossRef]
7. Jiang, J.; Rajendiran, S.; Piao, L.; Yoon, S. Base Effects on Carbonylative Polymerization of Propylene Oxide with a [(salph)Cr(THF)2]+[Co(CO)4](-) Catalyst. *Top. Catal.* **2017**, *60*, 750–754. [CrossRef]
8. Rajendiran, S.; Gunasekar, G.H.; Yoon, S. A heterogenized cobaltate catalyst on a bis-imidazolium-based covalent triazine framework for hydroesterification of epoxides. *New J. Chem.* **2018**, *42*, 12256–12262. [CrossRef]

9. Rajendiran, S.; Park, K.; Lee, K.; Yoon, S. Ionic-Liquid-Based Heterogeneous Covalent Triazine Framework Cobalt Catalyst for the Direct Synthesis of Methyl 3-Hydroxybutyrate from Propylene Oxide. *Inorg. Chem.* **2017**, *56*, 7270–7277. [CrossRef]
10. Kramer, J.W.; Rowley, J.M.; Coates, G.W. Ring-Expanding Carbonylation of Epoxides. In *Organic Reactions*; Denmark, S.E., Ed.; Wiley: Hoboken, NJ, USA, 2015; pp. 1–104.
11. Getzler, Y.D.Y.L.; Mahadevan, V.; Lobkovsky, E.B.; Coates, G.W. Synthesis of β-Lactones: A Highly Active and Selective Catalyst for Epoxide Carbonylation. *J. Am. Chem. Soc.* **2002**, *124*, 1174–1175. [CrossRef]
12. Kramer, J.W.; Lobkovsky, E.B.; Coates, G.W. Practical β-Lactone Synthesis: Epoxide Carbonylation at 1 atm. *Org. Lett.* **2006**, *8*, 3709–3712. [CrossRef] [PubMed]
13. Schmidt, J.A.R.; Mahadevan, V.; Getzler, Y.D.Y.L.; Coates, G.W. A Readily Synthesized and Highly Active Epoxide Carbonylation Catalyst Based on a Chromium Porphyrin Framework: Expanding the Range of Available β-Lactones. *Org. Lett.* **2004**, *6*, 373–376. [CrossRef] [PubMed]
14. Schmidt, J.A.R.; Lobkovsky, E.B.; Coates, G.W. Chromium (III) Octaethylporphyrinato Tetracarbonylcobaltate: A Highly Active, Selective, and Versatile Catalyst for Epoxide Carbonylation. *J. Am. Chem. Soc.* **2005**, *127*, 11426–11435. [CrossRef] [PubMed]
15. Park, H.D.; Dincă, M.; Román-Leshkov, Y. Heterogeneous Epoxide Carbonylation by Cooperative Ion-Pair Catalysis in Co(CO)4$^-$-Incorporated Cr-MIL-101. *ACS Central Sci.* **2017**, *3*, 444–448. [CrossRef]
16. Rajendiran, S.; Natarajan, P.; Yoon, S. A covalent triazine framework-based heterogenized Al-Co bimetallic catalyst for the ring-expansion carbonylation of epoxide to beta-lactone. *RSC Adv.* **2017**, *7*, 4635–4638. [CrossRef]
17. Jiang, J.; Yoon, S. A Metalated Porous Porphyrin Polymer with [Co(CO)4]$^-$ Anion as an Efficient Heterogeneous Catalyst for Ring Expanding Carbonylation. *Sci. Rep.* **2018**, *8*, 13243. [CrossRef]
18. Ganesan, V.; Yoon, S. Hyper-Cross-Linked Porous Porphyrin Aluminum(III) Tetracarbonylcobaltate as a Highly Active Heterogeneous Bimetallic Catalyst for the Ring-Expansion Carbonylation of Epoxides. *ACS Appl. Mater. Interfaces* **2019**, *11*, 18609–18616. [CrossRef]
19. Ganesan, V.; Yoon, S. Direct Heterogenization of Salphen Coordination Complexes to Porous Organic Polymers: Catalysts for Ring-Expansion Carbonylation of Epoxides. *Inorg. Chem.* **2020**, *59*, 2881–2889. [CrossRef]
20. Rajendiran, S.; Ganesan, V.; Yoon, S. Balancing between Heterogeneity and Reactivity in Porphyrin Chromium-Cobaltate Catalyzed Ring Expansion Carbonylation of Epoxide into β-Lactone. *Inorg. Chem.* **2019**, *58*, 3283–3289. [CrossRef]
21. Jiang, J.; Rajendiran, S.; Yoon, S. Double Ring-Expanding Carbonylation Using an In Situ Generated Aluminum Phthalocyanine Cobalt Carbonyl Complex. *Asian J. Org. Chem.* **2018**, *8*, 151–154. [CrossRef]
22. Rowley, J.M.; Lobkovsky, E.B.; Coates, G.W. Catalytic Double Carbonylation of Epoxides to Succinic Anhydrides: Catalyst Discovery, Reaction Scope, and Mechanism. *J. Am. Chem. Soc.* **2007**, *129*, 4948–4960. [CrossRef] [PubMed]
23. Church, T.L.; Getzler, Y.D.; Byrne, C.M.; Coates, G.W. Carbonylation of heterocycles by homogeneous catalysts. *Chem. Commun.* **2007**, *7*, 657–674. [CrossRef] [PubMed]
24. Chao, C.-G.; Bergbreiter, D.E. Highly organic phase soluble polyisobutylene-bound cobalt phthalocyanines as recyclable catalysts for nitroarene reduction. *Catal. Commun.* **2016**, *77*, 89–93. [CrossRef]
25. Ghani, F.; Kristen, J.; Riegler, H. Solubility Properties of Unsubstituted Metal Phthalocyanines in Different Types of Solvents. *J. Chem. Eng. Data* **2012**, *57*, 439–449. [CrossRef]
26. Bulgakov, B.A.; Sulimov, A.V.; Babkin, A.V.; Kepman, A.V.; Malakho, A.P.; Avdeev, V.V. Dual-curing thermosetting monomer containing both propargyl ether and phthalonitrile groups. *J. Appl. Polym. Sci.* **2017**, *134*, 44786. [CrossRef]
27. Lever, A. The Phthalocyanines. In *Advances in Inorganic Chemistry and Radiochemistry*; Emeléus, H.J., Sharpe, A.G., Eds.; Academic Press: Cambridge, MA, USA, 1965; Volume 7, pp. 27–114.
28. Lee, W.; Yuk, S.B.; Choi, J.; Jung, D.H.; Choi, S.-H.; Park, J.; Kim, J.P. Synthesis and characterization of solubility enhanced metal-free phthalocyanines for liquid crystal display black matrix of low dielectric constant. *Dye. Pigment.* **2012**, *92*, 942–948. [CrossRef]
29. Padmanaban, S.; Yoon, S. Surface Modification of a MOF-based Catalyst with Lewis Metal Salts for Improved Catalytic Activity in the Fixation of CO_2 into Polymers. *Catalysts* **2019**, *9*, 892. [CrossRef]

30. Donohue, M.; Aranovich, G. A new classification of isotherms for Gibbs adsorption of gases on solids. *Fluid Phase Equilibria* **1999**, *158*, 557–563. [CrossRef]
31. He, W.-L.; Wu, C. Incorporation of Fe-phthalocyanines into a porous organic framework for highly efficient photocatalytic oxidation of arylalkanes. *Appl. Catal. B: Environ.* **2018**, *234*, 290–295. [CrossRef]
32. McKeown, N.B.; Makhseed, S.; Budd, P.M. Phthalocyanine-based nanoporous network polymers. *Chem. Commun.* **2002**, 2780–2781. [CrossRef]
33. Maffei, A.V.; Budd, P.M.; McKeown, N.B. Adsorption Studies of a Microporous Phthalocyanine Network Polymer. *Langmuir* **2006**, *22*, 4225–4229. [CrossRef] [PubMed]
34. Neti, V.S.P.K.; Wang, J.; Deng, S.; Echegoyen, L. High and selective CO_2 adsorption by a phthalocyanine nanoporous polymer. *J. Mater. Chem. A* **2015**, *3*, 10284–10288. [CrossRef]
35. Ma, P.; Lv, L.; Zhang, M.; Yuan, Q.; Cao, J.; Zhu, C. Synthesis of catalytically active porous organic polymer from iron phthalocyanine and diimide building blocks. *J. Porous Mater.* **2015**, *22*, 1567–1571. [CrossRef]
36. Xue, Q.; Xu, Z.; Jia, D.; Li, X.; Zhang, M.; Bai, J.; Li, W.; Zhang, W.; Zhou, B.; Wang, J. Solid-Phase Synthesis Porous Organic Polymer as Precursor for Fe/Fe_3C-Embedded Hollow Nanoporous Carbon for Alkaline Oxygen Reduction Reaction. *ChemElectroChem* **2019**, *6*, 4491–4496. [CrossRef]
37. Ganji, P.; Doyle, D.J.; Ibrahim, H. In situ Generation of the Coates Catalyst: A Practical and Versatile Catalytic System for the Carbonylation of meso-Epoxides. *Org. Lett.* **2011**, *42*, 3142–3145. [CrossRef] [PubMed]
38. Kim, M.H.; Song, T.; Seo, U.R.; Park, J.E.; Cho, K.; Lee, S.M.; Kim, H.; Ko, Y.-J.; Chung, Y.K.; Son, S.U. Hollow and microporous catalysts bearing Cr(III)–F porphyrins for room temperature CO_2 fixation to cyclic carbonates. *J. Mater. Chem. A* **2017**, *5*, 23612–23619. [CrossRef]
39. Deng, F.-G.; Hu, B.; Sun, W.; Chen, J.; Xia, C. Novel pyridinium based cobalt carbonyl ionic liquids: Synthesis, full characterization, crystal structure and application in catalysis. *Dalton Trans.* **2007**, 4262–4267. [CrossRef]
40. Edgell, W.F.; Lyford, J. Preparation of sodium cobalt tetracarbonyl. *Inorg. Chem.* **1970**, *9*, 1932–1933. [CrossRef]

© 2020 by the authors. Licensee MDPI, Basel, Switzerland. This article is an open access article distributed under the terms and conditions of the Creative Commons Attribution (CC BY) license (http://creativecommons.org/licenses/by/4.0/).

Article

Flow Pd(II)-Catalysed Carbonylative Cyclisation in the Total Synthesis of Jaspine B

Pavol Lopatka [1], Michal Gavenda [1], Martin Markovič [1,2,*], Peter Koóš [1,2,*] and Tibor Gracza [1]

[1] Institute of Organic Chemistry, Catalysis and Petrochemistry, Slovak University of Technology, Radlinského 9, 812 37 Bratislava, Slovakia; pavol.lopatka@stuba.sk (P.L.); michal.gavenda@stuba.sk (M.G.); tibor.gracza@stuba.sk (T.G.)
[2] Georganics, Ltd., Koreničova 1, 811 03 Bratislava, Slovakia
* Correspondence: martin.markovic@stuba.sk (M.M.); peter.koos@stuba.sk (P.K.); Tel.: +421-259-325-129 (P.K.)

Abstract: This work describes the total synthesis of jaspine B involving the highly diastereoselective Pd(II)-catalysed carbonylative cyclisation in the preparation of crucial intermediates. New conditions for this transformation were developed and involved the pBQ/LiCl as a reoxidation system and $Fe(CO)_5$ as an in situ source of stoichiometric amount of carbon monoxide (1.5 molar equivalent). In addition, we have demonstrated the use of a flow reactor adopting proposed conditions in the large-scale preparation of key lactones.

Keywords: Jaspine B; flow chemistry; palladium catalysis; cyclisation; carbonylation

---

Citation: Lopatka, P.; Gavenda, M.; Markovič, M.; Koóš, P.; Gracza, T. Flow Pd(II)-Catalysed Carbonylative Cyclisation in the Total Synthesis of Jaspine B. Catalysts **2021**, 11, 1513. https://doi.org/10.3390/catal11121513

Academic Editor: Laura Antonella Aronica

Received: 26 November 2021
Accepted: 9 December 2021
Published: 12 December 2021

Publisher's Note: MDPI stays neutral with regard to jurisdictional claims in published maps and institutional affiliations.

Copyright: © 2021 by the authors. Licensee MDPI, Basel, Switzerland. This article is an open access article distributed under the terms and conditions of the Creative Commons Attribution (CC BY) license (https://creativecommons.org/licenses/by/4.0/).

1. Introduction

The examination of natural resources clearly remains the basis of the discovery of new bioactive substances. Since these newly discovered natural compounds become the inspiration for novel drug candidates, many groups in the scientific community create their research programs aiming at these novel structures. As a result, the newly developed transformations are then presented in terms of their applicability in the synthesis of such targets. However, in many cases such synthetic demonstrations do not provide accessibility to all derivatives and/or usable amounts of promising target molecule for further testing. In particular, progress in later stages of pharmaceutical/biomedical research might then be negatively affected. Although the synthetic optimisation is not so scientifically valued in the chemical community, it is still of great importance. This is especially so today when we are facing environmental and climate changes and need to be concerned about a sustainable future.

Since its discovery, jaspine B (pachastrissamine) **1** has drawn an immense attention from the scientific community (Figure 1).

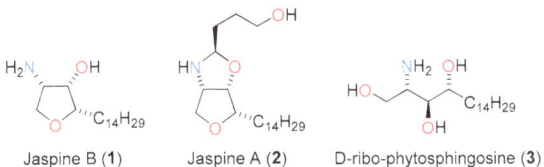

Jaspine B (**1**)   Jaspine A (**2**)   D-ribo-phytosphingosine (**3**)

Figure 1. Structure of pachastrissamine (jaspine B) (**1**), jaspine A (**2**) a D-ribophytosphingosine (**3**).

This natural sphingolipid **1** was independently isolated in 2002 and 2003 by T. Higa et al. and by the C. Debitus group from marine sponges, *Pachastrissa* sp. (family Calthropellidae) and *Jaspis* sp., respectively [1,2]. Due to its similar structure to other bioactive sphingolipids, jaspine B **1** has been also tested for its pharmacological properties and it has shown in vitro cytotoxic activity against several types of cancer cell (A-549, P-388, HT-29,

MeL-28, MCF-7, KB, HCT-116, U2OS, MDA-231. HeLa, CNE, MGC-803, EC-9706, PC-3, A-375, WM-115, Caco-2, Jurkat, SNU-638 and Caki-1) in the micro and sub-micromolar range [1–15]. The group of Y. Salma described how the cytotoxic activity of jaspine B **1** is based on the inhibition of SMS (sphinghomyelin synthetase) enzyme activity, which is responsible for maintaining stable concentration levels of ceramides (Cer) in the cell. Thus, higher concentration of Cer induces the cell apoptosis by a caspase-dependent pathway.

To this date, an enormous effort has been made to prepare and study this natural compound **1**. There are 35 known syntheses of jaspine B **1**, ten of which are based on an asymmetric step [12,16–24] and twenty-five are chiral pool approaches [4,8,9,25–45]. Moreover, the promising biological activity of this natural compound has resulted in the syntheses of various derivatives of this molecule for structure–activity relationship study purposes (Figure 2).

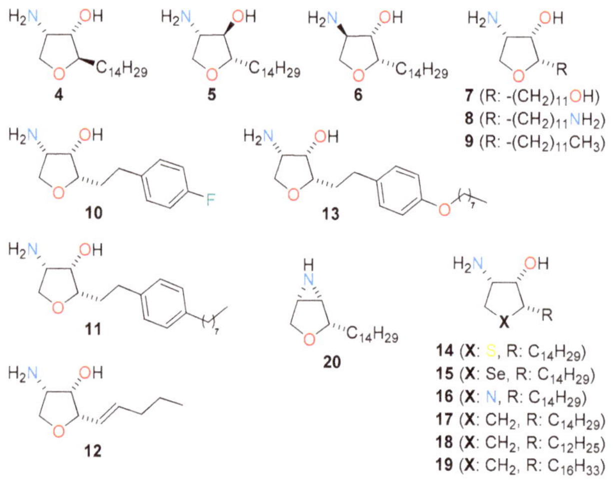

Figure 2. Synthetic derivatives of jaspine B **1**.

In summary, the biological activity of **1** has been found to be highly dependent on the stereo-configuration of the ring substituents and on the length of the aliphatic chain of the natural product [4,9]. The best bioactivity was observed while keeping the original configuration and the length of aliphatic chain [31]. However, the oxygen atom in the heterocycle of jaspine B **1** has not been found to be crucial for its cytotoxic properties [46].

2. Results and Discussion

In the course of our long-term research program directed towards CO gas-free carbonylative cyclisations, flow transformations and their synthetic applications, we have developed a new flow protocol for Pd-catalysed carbonylation reactions based on the use of iron pentacarbonyl [47]. To this date, only a few flow applications of this stereoselective reaction are reported in the literature. However, many total syntheses of natural products utilised this transformation as a batch process [48–50]. As a result, the reaction conditions have undergone many changes and the reoxidation system as one of the most modified parameters has been varied/adjusted from its original conditions to substrate specific requirements (pBQ/CuCl$_2$/O$_2$) [51,52].

Based on the previous results and our experience with the Pd-catalysed carbonylation in the total synthesis of natural products, we have proposed a total synthesis of jaspine B **1** utilising the flow carbonylative step as a key transformation.

2.1. Total Synthesis of Jaspine B

The synthesis of jaspine B **1** was designed to build the chiral centres via the stereoselective Pd-catalysed carbonylation. In addition, the synthesis was optimised to reach one of the key aspects—the compatibility of batch reaction conditions for the following application in the flow system. Thus, this key transformation would provide N-protected lactone **21** with correct configuration at chiral centres. The substrate for this cyclisation—unsaturated N-protected amino diol can be easily accessible from L-serine (Scheme 1).

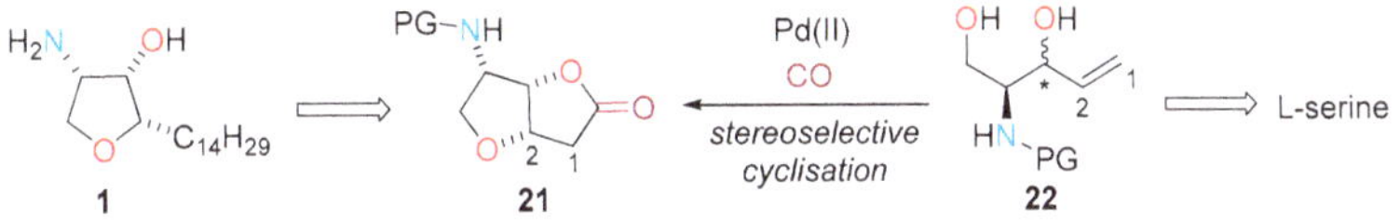

Scheme 1. Retrosynthetic analysis of jaspine B **1**.

As depicted in the retrosynthetic analysis, further transformation (chain elongation) of lactone **21** functional group would lead to the natural jaspine B.

Accordingly, the synthesis of **1** started from a commercially available N-Boc (N-t-butoxycarbonyl) protected Garner's aldehyde **26**-Boc. This aldehyde **26**-Boc can be easily prepared in few synthetic steps starting from L-serine. The whole sequence involving an esterification of L-serine, amino group protection of derivate **23**, the formation of oxazolidine **25** and the following reduction of the ester group is very well described in the literature [53] (Scheme 2).

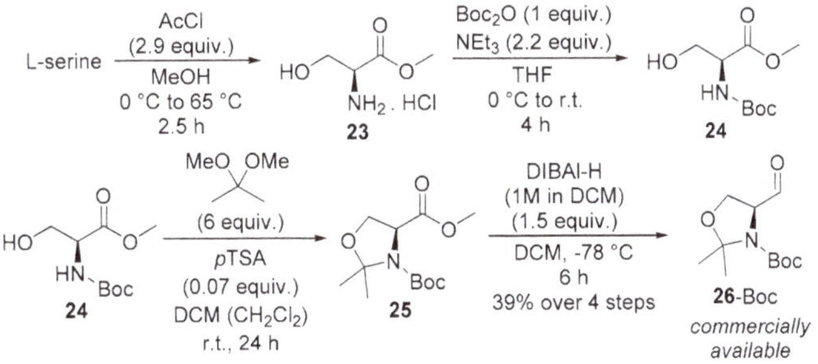

Scheme 2. Preparation of N-Boc protected Garner's aldehyde **27**-Boc.

At the beginning, the N-Boc protected Garner's aldehyde **26**-Boc was transformed to unsaturated N-Boc protected aminodiol **22**-Boc by a two step sequence (Scheme 3a) [54]. In detail, the addition of vinylmagnesium bromide to the starting aldehyde **26**-Boc in THF (tetrahydrofuran) provided a mixture of diastereomeric alcohols **27**-Boc in 91% yield. The selectivity of this reaction varies from 90:10 to 60:40 depending on the reaction temperature [54–56]. The major isomer, 2S,3R-alcohol **27**-Boc leads to a final product **1** with correct stereo configuration. The following selective deprotection of the acid labile oxazolidine group of **27**-Boc using *p*TSA (*p*-toluenesulfonic acid) provided N-Boc protected aminodiols **22**-Boc in 76% yield as an inseparable mixture [57]. Such a mixture of diastereomeric alcohols was then submitted to Pd-catalysed carbonylative cyclisation. This reaction was performed using well-established conditions with $Fe(CO)_5$ as a CO surrogate and the desired products-lactones **21**-Boc were obtained in 75% combined yield. At this stage, the lactone **21**-Boc with all *syn* configuration was separated using MPLC as a major substance from the diastereomeric mixture. Next, the reduction of lactone functional group using DIBAL-H (diisobutylaluminum hydride) provided lactol **28**-Boc in 88% yield (Scheme 3b).

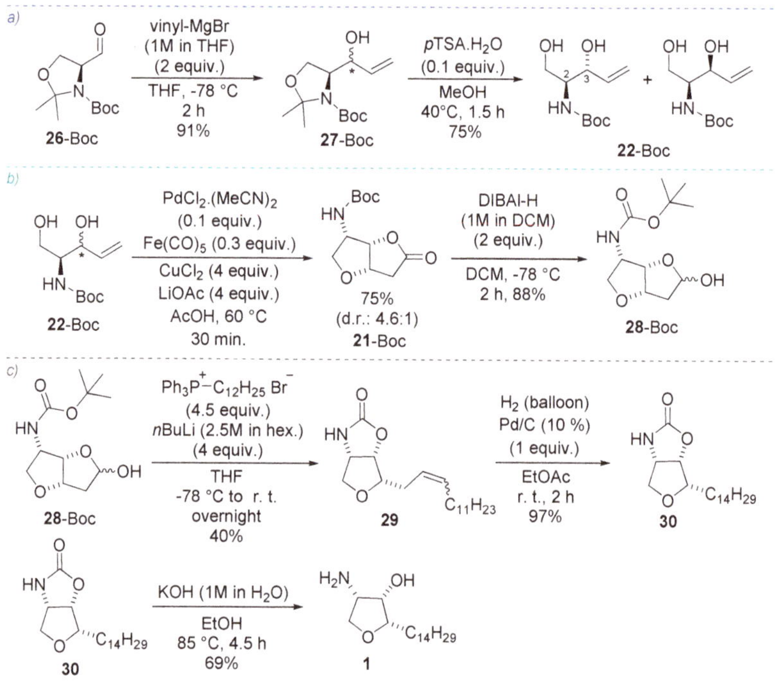

Scheme 3. Total synthesis of jaspine B **1**.

The final step sequence included a Wittig reaction, double bond reduction and oxazolidinone cleavage (Scheme 3c). These synthetic steps have already been described in the literature and the authors used them to furnish the final natural compound **1** by employing N-Cbz (N-benzyloxycarbonyl) protected lactone **21-Cbz** as a starting material [35]. However, based on the inspection of spectroscopic data, Davies et al. [58] later described the epimerization at C-2 carbon occurred during the Wittig reaction and the authors prepared 2-*epi*-jaspine B [35]. This discrepancy was accounted for by a retro-Michael/Michael epimerisation reaction pathway upon treatment of lactol **28** with excess Wittig reagent (Scheme 4b).

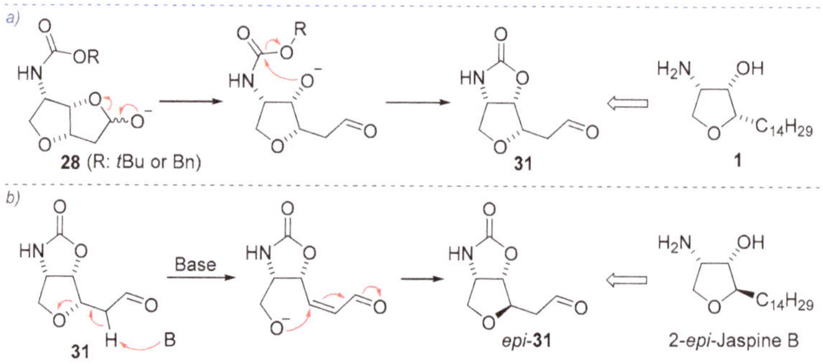

Scheme 4. Formation of oxazolidinone ring (**a**) and possible epimerisation at C-2 carbon in Wittig reaction (**b**).

The authors later described how this epimerisation occurred in the DIBAL-H reduction step using old reagent containing base via same reaction pathway [15].

In our case, by applying the described conditions we were able to prepare the chiral oxazolidinone **29**-Boc in 40% yield. The formation of oxazolidinone ring via intramolecular attack of the hydroxy anion to the carbamate group furnished aldehyde **31** (Scheme 4a). Following a reduction of the double bond using Pd/C and H$_2$ gave compound **30** in 97% yield. Finally, jaspine B **1** was provided by a cleavage of the oxazolidinone ring of **30** using aqueous KOH in 69% yield. The comparison of spectroscopic data of prepared jaspine B **1** to described data [34] revealed that the epimerisation of **31** during Wittig reaction or reduction has not occurred.

In summary, we have accomplished the total synthesis of jaspine B **1** in seven steps starting from N-Boc protected Garner's aldehyde. The stereoselective Pd(II)-catalysed carbonylative cyclisation was used in the preparation of key intermediate-lactone **21**-Boc. The yield of jaspine B was 10% over all synthetic steps.

In addition, the synthesis of jaspine B was also performed starting from the commercially available N-Cbz protected Garner's aldehyde **26**-Cbz. Following the same synthetic route, we were able to increase the overall yield of jaspine B up to 16% yield over seven reaction steps (Scheme 5).

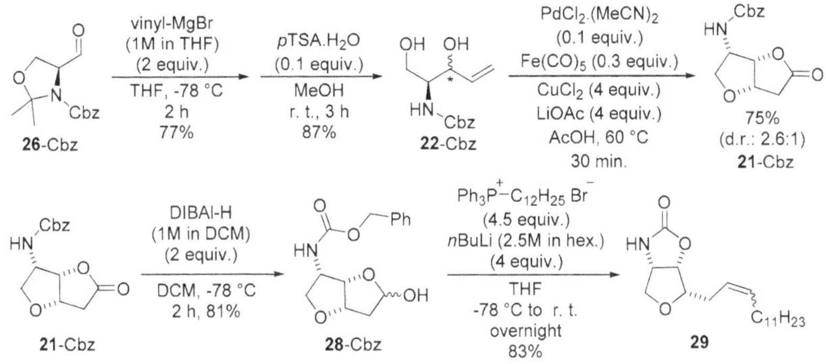

Scheme 5. Synthesis of jaspine B **1** via N-Cbz protected lactone.

Similarly, the Wittig reaction of lactol **28**-Cbz proceeded cleanly to oxazolidinone **29** without unwanted epimerisation at the C-2 carbon centre as in the previous case and the desired product **29** was isolated in 83% yield.

After successful optimisation of the total synthesis of jaspine B in batch, we focused our attention on the application of flow chemistry for the preparation of key intermediate-lactone **21**. Thus, the flow Pd(II)-catalysed carbonylative cyclisation of **22** was proposed.

2.2. Flow Synthesis of the Key Intermediate 21 for the Preparation of Jaspine B 1

Over the last few decades, the flow chemistry has shown many advantages in organic synthesis, and it has grown into a modern and enabling tool for new synthetic methods utilising dangerous and/or toxic chemicals [59,60]. Moreover, many total syntheses of various biorelevant compounds have utilised this technique as a fundamental instrument in the preparation of key intermediates [61–63] or as a multiple step telescoped system [64–66]. At present, the flow chemistry has become an integral part of scientific research and it is commonly used in synthetic laboratories at universities and in pharmaceutical companies.

The flow chemistry technique as a part of our research program has been used in the application of CO surrogates in the carbonylative transformations. We have previously demonstrated that Fe(CO)$_5$ can be utilised as a CO donor in Pd(II)-catalysed carbonylative cyclisation [47]. In the continuation of our research, we have focused on the development of new conditions for this flow transformation and on its application in the flow synthesis of bioactive compounds. Thus, we have proposed a new flow system for the preparation of the key intermediate **21** in the jaspine B **1** synthesis (Figure 3).

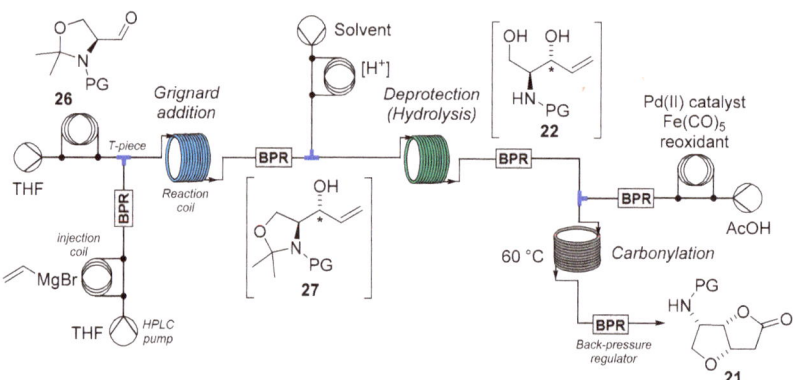

Figure 3. Proposed telescoped flow system for the synthesis of bicyclic lactone **21**.

The proposed telescoped flow system consists of three parts involving the Grignard addition, selective deprotection and crucial cyclisation. The aim of the design was to perform these steps in a one-telescoped system without executing any isolation procedure. Consequently, the flow system would provide the crude final lactone **21** starting from commercially available substrates.

Our investigation started with a series of batch experiments aimed to address the minimal requirements and compatibility of reaction conditions for flow procedure. At first, we examined various reaction conditions of the first two steps sequence of the total synthesis—addition of vinyl magnesium bromide to N-protected Garner's aldehyde **26** and selective cleavage of oxazolidine **27** without the isolation after the first step (Scheme 6).

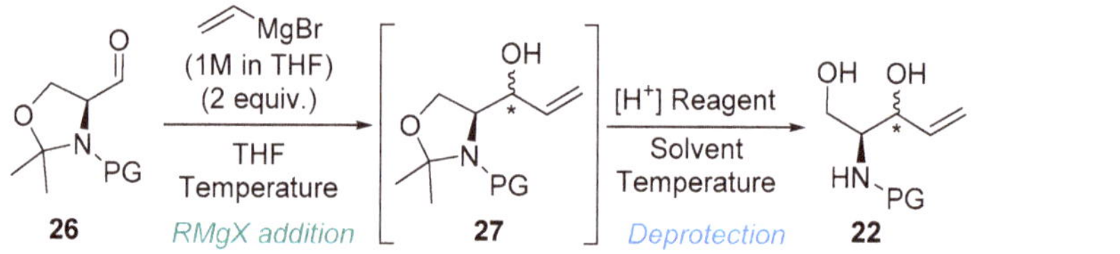

Scheme 6. Optimisation of the RMgX addition and deprotection reaction sequence.

Compared to the previously described batch conditions (Scheme 3a), we tried and modified mainly the cleavage of 2,2-dimethyl oxazolidine **27** using different acidic conditions (Table 1). The best results were achieved using pTSA.H$_2$O as a H$^+$ source in MeOH in the second step (Table 1, Entry 1 and 5). Following products **22**-Cbz and **22**-Boc were isolated in 87% and 75% yield over two steps, respectively. In general, altering the temperature of RMgX addition did not affect the yield of this sequence and only the difference in the diastereoselectivity outcome was observed. The addition step at 0 °C of this two-step procedures provided in all cases a 1:1 mixture of diastereomeric alcohols **22**. In addition, the SO$_3$H polymer supported resin, Amberlyst 15 was also tried for the deprotection step as the implementation of polymer supported reagents in flow reactions is described very well [64,67]. In this case, the yield of product **22**-Boc was decreased due to the partial cleavage of an acid labile N-Boc protecting group of substrate **26**-Boc. The formation of completely deprotected aminodiol **22** was confirmed by LC-MS analysis and the full cleavage of protecting groups of **26**-Boc was also observed in the case of the reaction performed in AcOH (Table 1, entry 3).

Table 1. Optimisation of the RMgX addition and deprotection reaction sequence in batch.

Entry	Substrate	RMgX Addition			Deprotection				Product
		T (°C)	Rxn Time	[H⁺] Reagent (Equivalent)	Solvent	T (°C)	Rxn Time	Yield (%) [a]
1	26-Boc	−78	2 h	pTSA.H₂O (0.1)	MeOH	40 °C	1.5 h	75
2	26-Boc	0	2 h	Amberlyst 15 (3.2) [b]	MeOH	40 °C	0.5 h	46 [c]
3	26-Boc	0	2 h	-	AcOH	70 °C	0.5 h	35 [c]
4	26-Cbz	−78	1.5 h	-	AcOH/H₂O (4/1)	r.t.	3 h	59
5	26-Cbz	−78	1.5 h	pTSA.H₂O (0.1)	MeOH	r.t.	3 h	87
6	26-Cbz	0	1.5 h	Amberlyst 15 (3.2) [b]	MeOH	40 °C	0.5 h	73

[a] Yield based on the amount of isolated product **22**. [b] The capacity of Amberlyst 15 is 4.7 m equivalents per 1g by dry weight. [c] Partial cleavage of *N*-Boc protecting group was observed by LC-MS analysis.

Based on the optimisation of reaction sequence under batch conditions, we performed a series of experiments using a flow system as depicted in Scheme 7. At first, only nucleophilic addition of RMgX was examined. In this case, the second stage of the flow setup was omitted, and the Grignard reaction was performed at 0 °C to ensure the homogeneity of the reaction stream (the reaction at −78 °C is not homogeneous). Thus, the flow reaction using a 9 mL reactor coil provided products **27** with full conversion of substrates **26** in acceptable yields with lower stereoselectivity (Table 2, entry 1 and 2).

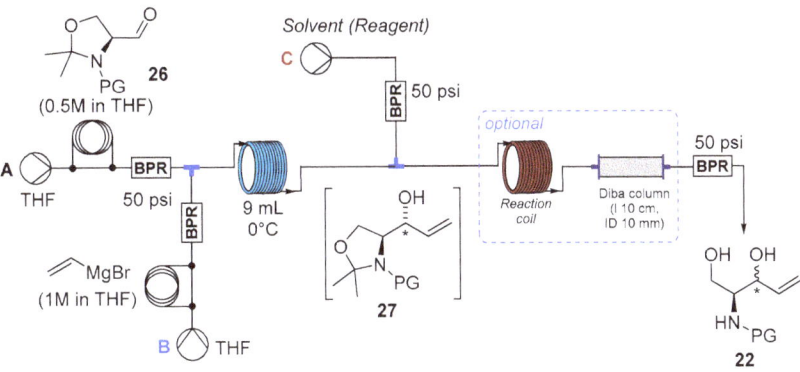

Scheme 7. Optimisation of the flow addition/deprotection reaction sequence.

The following flow experiments employing the second stage of the system were performed using a reaction coil or Diba column (optional) depending on the use of H⁺ donor in the acetonide deprotection step. In the case of the reaction using an excess of Amberlyst 15 (1.7 g, 8 equivalents) in the Diba column, the reagent also secured the filtration of reaction stream and Mg(II) salts formed after mixing (quenching) the Grignard reaction stream with MeOH were caught on the polymer resin. However, the larger excess of Amberlyst 15 also caused a parallel carbamate cleavage and decreased the yield of products. Thus, the flow system using Amberlyst 15 provided products **22**-Boc and **22**-Cbz in 34 and 49% yield, respectively (Table 2, entry 3 and 4).

The best results were achieved using 1.1 M solution of pTSA in MeOH in the second step and the products were isolated in similar yields to batch (Table 2, entry 5 and 6). In this case, the concentration and flow rate of *para*-toluenesulfonic acid stream were adjusted to ensure the catalytic amount of H⁺ necessary for the deprotection of acetonide group.

Since, the 0.25 of 0.275 mmol/min amount of this acid was immediately quenched in the reactor by the excess of Mg(II) salts, the 0.275 mmol/min amount really represents only the catalytic amount of H^+. Thus, the flow reactions using 8.65 mmol of substrates **27-Cbz** or **27-Boc** led to the formation of desired N-protected unsaturated aminodiols **22** in 82 and 62% yields, respectively (Table 2, entry 5 and 6). With optimised conditions for this addition/deprotection sequence in hand, we turned our attention to the continuous carbonylative cyclisation step.

Table 2. Optimisation of the flow addition/deprotection reaction sequence.

Entry	Substrate 26-PG	STREAM						Optional Equipment	Yield [b] (%)
		A [a]		B		C			
		Inj. Coil (mL)	Flow Rate (mL/min.)	Inj. Coil (mL)	Flow Rate (mL/min.)	Solvent	Flow Rate (mL/min.)		
1	Boc	2	0.3	2.3	0.3	-	-		84 [c]
2	Cbz	2	0.25	2.3	0.25	-	-		66 [c]
3	Boc	2	0.3	2.3	0.3	MeOH	0.6	1.7g (Amberlyst)	34 [d]
4	Cbz	2	0.25	2.3	0.25	MeOH	0.5	1.7g (Amberlyst)	49
5	Boc	17.3	0.25	18.4	0.25	pTSA in MeOH (1.1M)	0.25	reaction coil (10 mL, 50 °C)	62
6	Cbz	17.3	0.25	18.4	0.25	pTSA in MeOH (1.1M)	0.25	reaction coil (10 mL, 50 °C)	82

[a] 0.5 M solution of substrate in THF (tetrahydrofuran) was used. [b] Yield based on the amount of isolated product **22**. [c] Only addition step was performed in the flow system, the yield corresponds to the addition product **27**. [d] Deprotection of N-Boc group was also observed.

In 2018, we reported on pros and cons of the flow and batch stereoselective Pd-catalysed carbonylative cyclisation of unsaturated polyols/aminoalcohols using known conditions [68] and $Fe(CO)_5$ as an in situ donor of carbon monoxide [47]. By adjusting the concentration of reaction streams and the amount of Cu^{2+}/Li^+ inorganic salts required for the reoxidation Pd^0, we were able to prepare a series of various cyclisation products in a flow system (Scheme 8).

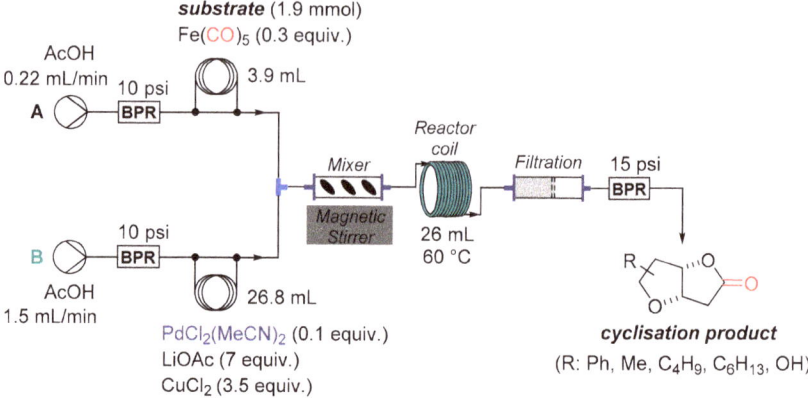

Scheme 8. Previously described continuous Pd(II)-catalysed carbonylation of unsaturated polyols and aminoalcohols.

Even though aforementioned continuous flow system was successfully applied in the large-scale preparation of desired bicyclic lactones, there are still limitations of the system regarding the formation of insoluble copper salts in the reactor. Also, an active mixer and interchangeable filtration unit were necessary to perform the transformation over longer periods.

With the aim to improve the conditions for continuous flow Pd-catalysed oxy/aminocarbonylation, we adopted the use of *p*BQ (*p*-benzoquinone) instead of Cu^{2+} as a reoxidant.

At first, we performed a batch reaction using 2 equivalents of *p*BQ, 0.6 equivalent of $Fe(CO)_5$ and 0.1 equivalent $PdCl_2(MeCN)_2$ using diastereomeric mixture **22**-Boc (0.23 mmol) in acetic acid (0.9 mL, 0.25 M reaction). The reaction at 60 °C proceeded with full conversion of starting material after 1 h, however noticeable amounts of insoluble material were observed. The homogeneity of the reaction mixture over the whole course of reaction was achieved by the addition of 1 equivalent of LiCl. The reaction proceeded smoothly with full conversion of **22**-Boc in 1 h at 60 °C. A comparison of newly optimised and previously described conditions is shown in Figure 4.

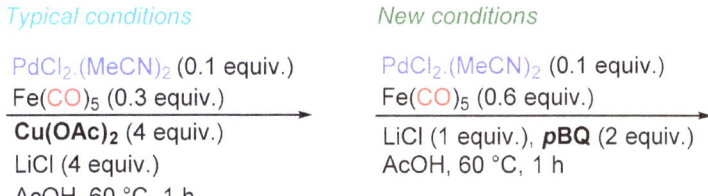

Figure 4. Comparison of known and new conditions for Pd-catalysed carbonylative cyclisation.

The new reaction conditions were then tested in the preparation of key intermediate-lactone **21** for the synthesis of jaspine B **1** in continuous flow mode (Scheme 9). The flow setup consisted of two reaction streams which were pumped using HPLC Azura pumps via injection coils to a preheated reactor coil. The composition of stock solutions were adjusted to avoid the decomposition of $Fe(CO)_5$ in the presence of oxidation reagents (*p*BQ). The following continuous reaction of **22** on 0.5 mmol scale provided the desired N-protected bicyclic lactones **21** in comparable yields to standard batch and flow conditions (Table 3).

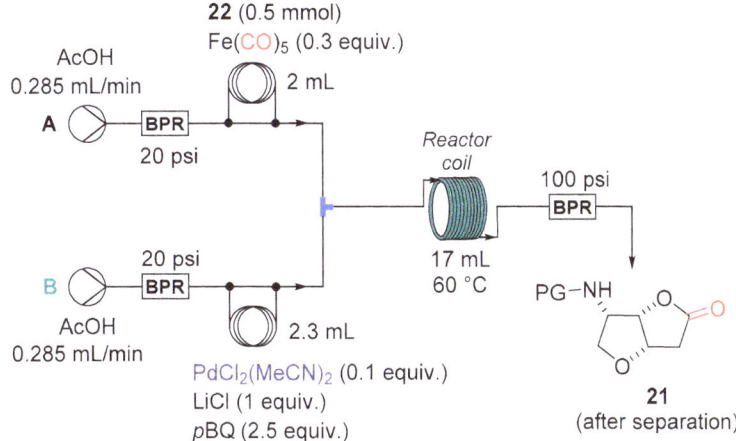

Scheme 9. Continuous Pd(II)-catalysed carbonylations of unsaturated N-protected aminodiols **22**.

Table 3. Comparison of Pd(II)-catalysed carbonylation of **22** employing batch and flow conditions.

Entry	Product	Typical Conditions		New Conditions
		Batch Yield [a,b] (%)	Flow Yield [a,c] (%)	Flow Yield [a,d] (%)
1	**21**-Cbz	75	76	71
2	**21**-Boc	75	64	72

[a] Combined yield of both diastereomers based on the isolated amounts of products **21**. [b] Batch reaction of **22** (0.5 mmol) using PdCl$_2$·2CH$_3$CN (0.048 mmol, 0.1 equivalent), CuCl$_2$ (1.9 mmol, 4 equivalents), LiOAc (1.9 mmol, 4 equivalents) and Fe(CO)$_5$ (0.15 mmol, 0.3 equivalent) in 1.9 mL of AcOH [68]. [c] Reaction was performed on 1.22 mmol scale (substrate **22**) using the conditions and flow system as described in the literature [47]. [d] Reaction was performed using flow system as depicted in Scheme 9.

In detail, flow transformation using new conditions as depicted in Scheme 9 provided both lactones **21** in comparable yields to reactions performed under typical conditions (Table 3, flow yield column). The new reoxidation system for Pd^0-Pd^{II} cycle ensures homogeneity of the reaction stream thus enabling better scaling of this flow transformation. Compared to the batch reaction (Table 3, typical conditions, batch yield column), the designed flow transformation (under new conditions) has several advantages. The batch reaction using typical conditions [47] can be undertaken only on a small scale due to the excessive pressure in the glass reaction tube. Upscaling the batch reaction 20 mmol may cause a few problems. As Fe(CO)$_5$ immediately decomposes after contact with the reaction mixture, it releases 1 equivalent of CO resulting in foaming and problematic stirring of heterogenous reaction mixture. Also, the pressure in a 120 mL reaction tube can raise up to 95 psi after few minutes.

To prove the robustness of the described flow setup, a six-hour long experiment was performed. Based on our previous experience with this type of transformation, we lowered the reaction stream concentration (0.25 M down to 0.125 M) to avoid the gas-liquid segment formation in the reaction coil. Continuous flow transformation of diastereomeric mixture **22**-Boc (5.57 g) using such minimally modified conditions provided the desired lactones **21**-Boc in 71% (4.5 g) combined yield (using the same flow setup as depicted in Scheme 10). However, a formation of precipitate at the exit of the reactor after cooling the reaction stream was observed. To prevent the potential clogging of the tubing, THF was employed as co-solvent in the case of a large-scale continuous reaction using **22**-Cbz (Scheme 10). In this case, the prepared stock solutions of substrate **22**-Cbz and reagents were pumped directly through the HPLC pumps into the larger reactor (47 mL) therefore allowing us to use higher flow rates (0.785 mL/min) and transform a larger amount of starting material **22**-Cbz in shorter time. In detail, the diastereomeric mixture N-Cbz protected aminodiols **22**-Cbz (7.6 g) was easily transformed over 2.5 h into bicyclic lactones **21**-Cbz (d.r.: 2.6:1, 6.3 g) in 75% combined yield. The pure diastereomer **21**-Cbz with all *syn*-configuration was obtained after MPLC purification in 54% yield (4.5 g).

In conclusion, we have designed and optimised an enhanced synthesis of the key intermediate-lactones **21**-Cbz and **21**-Boc utilising the stereoselective Pd-catalysed cyclo-carbonylation of corresponding unsaturated aminodiols **22**. The key lactones **21** were then successfully transformed into natural jaspine B **1** over a four-step sequence in batch. Also, we have demonstrated the applicability of the flow reactor in two steps preparation of N-protected aminodiols **22** in comparable yields to the batch process. Importantly, new conditions for Pd-catalysed cyclocarbonylation of unsaturated polyols/aminoacohols were developed involving *p*BQ/LiCl as a reoxidation system and Fe(CO)$_5$ as an in situ source of stochiometric amount of carbon monoxide (only 1.5 molar equivalents). Such conditions were easily applied to continuous flow mode allowing us to prepare gram quantities of intermediates **21** for jaspine B **1** synthesis. This flow setup has shown several advantages compared to previous versions of the flow reaction system and the homogeneity of the reaction stream facilitated the use of a common flow system without the implementation of any other special devices.

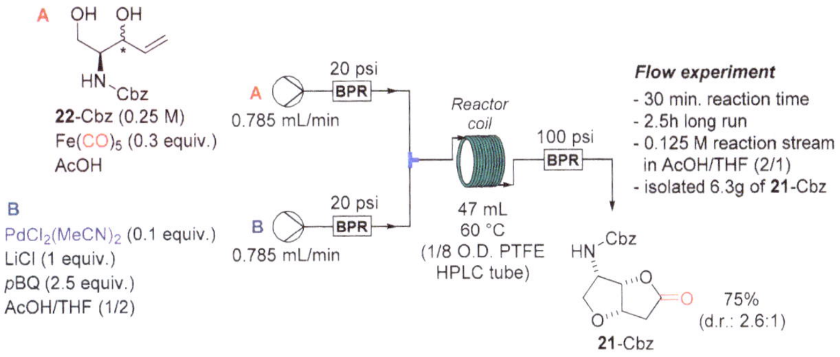

Scheme 10. Large-scale continuous synthesis of N-Cbz protected lactone **21-Cbz**.

3. Experimental Section

3.1. Material and Methods

Commercial materials which were obtained from Sigma-Aldrich, Acros Organics, Alfa Aesar or Fisher Scientific were used without further purification. Reactions were monitored using TLC on silica gel. Compound purification was undertaken by flash chromatography. All solvents were distilled before use. Hexanes refer to the fraction boiling at 60–65 °C. Flash column liquid chromatography (FLC) was performed on silica gel Kieselgel 60 (15–40 µm, 230–400 mesh) and analytical thin-layer chromatography (TLC) was performed on aluminium plates pre-coated with either 0.2 mm (DC-Alufolien, Merck) or 0.25 mm silica gel 60 F254 (ALUGRAM® SIL G/UV254, Macherey–Nagel). Analysed compounds were visualized by UV fluorescence and by dipping the plates in an aqueous H_2SO_4 solution of cerium sulphate/ammonium molybdate followed by charring with a heat gun. Melting points were obtained using a Boecius apparatus and are uncorrected. ^1H and ^{13}C NMR spectra were recorded on either 300 (75) MHz MercuryPlus or 600 (151) MHz Unity Inova spectrometers from Varian (Supplementary Materials). Chemical shifts (δ) are quoted in ppm and are referenced to the tetramethylsilane (TMS), $CDCl_3$ or DMSO-d_6 as internal standard. FTIR spectra were obtained on a Nicolet 5700 spectrometer (Thermo Electron) equipped with a Smart Orbit (diamond crystal ATR) accessory, using the reflectance technique (400–4000 cm^{-1}). High-resolution mass spectra (HRMS) were recorded on an OrbitrapVelos mass spectrometer (Thermo Scientific, Waltham, MA, USA; Bremen, Germany) with a heated electrospray ionization (HESI) source. The mass spectrometer was operated with a full scan (50–2000 amu) in positive or negative FT mode (at a resolution of 100,000). The sample was dissolved in methanol and infused via syringe pump at a rate of 5 mL/min. The heated capillary was maintained at 275 °C with a source heater temperature of 50 °C and the sheath, auxiliary and sweep gases were at 10, 5 and 0 units, respectively. Source voltage was set to 3.5 kV.

3.2. Representative Flow Procedures

3.2.1. Grignard Reaction

The flow setup consisted of three HPLC pumps (Knauer Azura 4.1S with 10 mL pump head). The pumps were used to introduce a solution of substrate **26** (1 mmol, 0.5 M) in anhydrous THF (Feed A), a commercial solution of vinylmagnesium bromide (1.0 M in THF, Sigma-Aldrich, Feed B), and MeOH (Feed C). Injection loops (PTFE, 0.8 mm i.d., 1.6 mm o.d.; internal volume: 2.0 mL, Feed A, and 2.3 mL, Feed B) were used to deliver the two starting feeds. At start of the experiment, the whole reactor was flushed with an anhydrous THF (Feed A and Feed B) and MeOH (Feed C). Both solutions were loaded into their corresponding injection loops. Feed A and feed B were pumped from the injection loops and mixed in a T-shaped connector (PEEK) in a cooling bath (0 °C). The combined mixture passed through a coil reactor (PTFE, 0.8 mm i.d., 1.6 mm o.d.; internal volume: 9.0 mL) at 0 °C before the mixture was combined with MeOH (Feed C) in a T-shaped connector (PEEK) at the same temperature. Final reaction mixture left the system through Upchurch BPR (15 psi). The mixture was then collected in the flask, and evaporated in vacuo. The residue was purified by MPLC (mixture of hexanes and EtOAc) providing the desired alcohols **27**.

3.2.2. Preparation of Unsaturated Aminodiols

The flow setup consisted of three HPLC pumps (Knauer Azura 4.1S with 10 mL pump head). The pumps were used to introduce a solution of substrate **26** (0.5 M) into anhydrous THF (Feed A), a commercial solution of vinylmagnesium bromide (1.0 M in THF, Sigma-Aldrich, Feed B), and quenching solvent/mixture (MeOH, AcOH, mixture AcOH/H_2O or 1.1 M solution of *p*TSA in MeOH), Feed C). Injection loops (PTFE, 0.8 mm i.d., 1.6 mm o.d.; internal volume: 2.0 mL, Feed A, and 2.3 mL, Feed A) were used to deliver the starting two feeds. At the beginning of the experiment, the complete reactor setup was flushed with anhydrous THF (Feed A and Feed B) and corresponding solvent/mixture (according to the conditions in Table 2, Feed C). Both solutions were loaded into their corresponding injection loops. Feed A and feed B were pumped from the injection loops and mixed in a T-shaped connector (PEEK) in a cooling bath (0 °C). The combined mixture passed through a coil reactor (PTFE, 0.8 mm i.d., 1.6 mm o.d.; internal volume: 9.0 mL) at 0 °C before the mixture was combined with Feed C (corresponding solvent or mixture) in a T-shaped connector (PEEK) at the same temperature. The mixture was then pumped through a second coil reactor at 50 or 70 °C (PTFE, 1.5 mm i.d., 3.2 mm o.d.; internal volume: 4.0 or 10.0 mL) or glass Omnifit column at 40 °C (10 mm i.d. × 100 mm length) filled with corresponding amount of Amberlyst 15. At the end, the reaction mixture left the system through Upchurch BPR and it was collected in the flask. The solvent from collected crude material was concentrated in vacuo (if the *p*TSA was used, collected stream was at first quenched with saturated water solution of $NaHCO_3$ and extracted with EtOAc). The residue was purified by MPLC (mixture of hexanes and EtOAc) providing the desired alcohol **22**.

3.2.3. Carbonylative Cyclisation Using pBQ/LiCl Reoxidation System

The flow setup consisted of two HPLC pumps (Knauer Azura 4.1S with 10 mL pump head). These pumps were used to introduce a solution of substrate **22** (0.25 M) and iron pentacarbonyl (0.3 equivalent) in glacial AcOH (Feed A), and solution of *p*BQ (2.5 equivalents), LiCl (1 equivalent) and $PdCl_2(MeCN)_2$ (0.1 equivalent) in the solvent (glacial AcOH or THF/AcOH = 2:1, Feed B). Injection loops (PTFE, 0.8 mm i.d., 1.6 mm o.d.; internal volume: 2.0 mL, Feed A, and 2.3 mL, Feed B) were used to deliver the two feeds. At the beginning of the experiment, the complete reactor setup was flushed with glacial AcOH (Feed A) and corresponding solvent/mixture (Feed B). Both solutions were loaded into their corresponding injection loops. (Stock solutions were pumped directly via HPLC pumps in the case of long runs). Feed A and feed B were pumped from the injection loops and mixed in a T-shaped connector (PEEK). The combined mixture went through a reactor coil (PTFE, 0.8 mm i.d., 1.6 mm o.d.; internal volume: 17.1 or 47.1 mL) at 60 °C before the flow stream left the system through Upchurch BPR (100 psi). The whole reaction stream was collected in the flask, and evaporated in vacuo. The residue was purified by MPLC (mixture of hexanes and EtOAc) providing the appropriate bicyclic lactones **21**.

Supplementary Materials: The following are available online at https://www.mdpi.com/article/10.3390/catal11121513/s1. All experimental procedures for batch and flow transformations, copies of ^{1}H and ^{13}C NMR spectra for all prepared compounds are included.

Author Contributions: Experimental work, P.L., M.G. and M.M.; design of experiments M.M. and P.K.; writing—original draft preparation, writing—review and editing, P.K. and M.M.; supervision, M.M., P.K. and T.G.; project administration, P.K. and M.M.; funding acquisition, P.K. All authors have read and agreed to the published version of the manuscript.

Funding: This research work was funded by SLOVAK GRANT AGENCIES APVV and VEGA, grant number APVV-20-0105 and VEGA No. 1/0552/18 and VEGA No. 1/0766/20.

Data Availability Statement: The datasets supporting the conclusions of this article are included within the article and Supplementary Materials.

Acknowledgments: We acknowledge the SLOVAK GRANT AGENCIES APVV, VEGA (APVV-20-0105 and VEGA No. 1/0552/18 and VEGA No. 1/0766/20) and Georganics Ltd. for funding.

Conflicts of Interest: The authors declare no conflict of interest.

References

1. Koruda, I.; Musman, M.; Ohtani, I.I.; Ichiba, T.; Tanaka, J.; Gravalos, D.G.; Higa, T. Pachastrissamine, a Cytotoxic Anhydrophytosphingosine from a Marine Sponge, Pachastrissa sp. *J. Nat. Prod.* **2002**, *65*, 1505–1506. [CrossRef]
2. Ledroit, V.; Debitus, C.; Lavaud, C.; Massiot, G. Jaspines A and B: Two new cytotoxic sphingosine derivatives from the marine sponge Jaspis sp. *Tetrahedron Lett.* **2003**, *44*, 225–228. [CrossRef]
3. Canals, D.; Moemenoe, D.; Farias, G.; Llebaria, A.; Casa, J.; Delgado, A. Synthesis and biological properties of Pachastrissamine (jaspine B) and diastereoisomeric jaspines. *Bioorg. Med. Chem.* **2009**, *17*, 235–241. [CrossRef] [PubMed]
4. Ghosal, P.; Ajay, S.; Meena, S.; Sinha, S.; Shaw, A.K. Stereoselective total synthesis of jaspine B (pachastrissamine) utilizing iodocyclization and an investigation of its cytotoxic activity. *Tetrahedron Asymmetry* **2013**, *24*, 903–908. [CrossRef]
5. Salma, Y.; Lafont, E.; Therville, N.; Carpentier, S.; Bonnafé, M.-J.; Levade, T.; Génisson, Y.; Andrieu-Abadie, N. The natural marine anhydrophytosphingosine, Jaspine B, induces apoptosis in melanoma cells by interfering with ceramide metabolism. *Biochem. Pharmacol.* **2009**, *78*, 477–485. [CrossRef]
6. Jayachitra, G.; Sudhakar, N.; Anchoori, R.K.; Rao, B.V.; Roy, S.; Banerjee, R. Stereoselective synthesis and biological studies of the C2 and C3 epimer and the enantiomer of Pachastrissamine (Jaspine B). *Synthesis* **2010**, *1*, 115–119.
7. Xu, J.-M.; Zhang, E.; Shi, X.-J.; Wang, Y.-C.; Yu, B.; Jiao, W.-W.; Guo, Y.-Z.; Liu, H.-M. Synthesis and preliminary biological evaluation of 1,2,3-triazole-Jaspine B hybrids as potential cytotoxic agents. *Eur. J. Med. Chem.* **2014**, *80*, 593–604. [CrossRef]
8. Martinková, M.; Mezeiová, E.; Fabišíková, M.; Gonda, J.; Pilátová, M.; Mojžiš, J. Total synthesis of pachastrissamine together with its 4-epi-congener via [3,3]-sigmatropic rearrangements and antiproliferative/cytotoxic evaluation. *Carbohyde. Res.* **2015**, *402*, 6–24. [CrossRef]
9. Santos, C.; Fabing, I.; Saffon, N.; Ballereau, S.; Génisson, Y. Rapid access to jaspine B and its enantiomer. *Tetrahedron* **2013**, *69*, 7227–7233. [CrossRef]
10. Jeon, H.; Bae, H.; Baek, D.; Kwak, Y.-S.; Kim, D.; Kim, S. Syntheses of sulfur and selenium analogues of pachastrissamine via double displacements of cyclic sulfate. *Org. Biomol. Chem.* **2011**, *9*, 7237–7242. [CrossRef]
11. Rives, A.; Ladeira, S.; Levade, T.; Andrieu-Abadie, N.; Génisson, Y. Synthesis of Cytotoxic Aza Analogues of Jaspine B. *J. Org. Chem.* **2010**, *75*, 7920–7923. [CrossRef]
12. Salma, Y.; Ballereau, S.; Maaliki, C.; Ladeira, S.; Andrieu-Abadie, N.; Génisson, Y. Flexible and enantioselective access to jaspine B and biologically active chain-modified analogues thereof. *Org. Biomol. Chem.* **2010**, *8*, 3227–3243. [CrossRef]
13. Salma, Y.; Ballereau, S.; Ladeira, S.; Lepetit, C.; Chauvin, R.; Andrieu-Abadie, N. Single- and double-chained truncated jaspine B analogues: Asymmetric synthesis, biological evaluation and theoretical study of an unexpected 5-endo-dig process. *Tetrahedron* **2011**, *67*, 4253–4262. [CrossRef]
14. Ballereau, S.; Andrieu-Abadie, N.; Saffon, N.; Génisson, Z. Synthesis and biological evaluation of aziridine-containing analogs of phytosphingosine. *Tetrahedron* **2011**, *67*, 2570–2578. [CrossRef]
15. Pashikanti, A.; Ukani, R.; David, S.A.; Datta, A. Total Synthesis and Structure–Activity Relationship Studies of the Cytotoxic Anhydrophytosphingosine Jaspine B (Pachastrissamine). *Synthesis* **2017**, *49*, 2088–2100.
16. Schmiedel, V.M.; Stefani, S.; Reissig, H.-U. Stereodivergent synthesis of jaspine B and its isomers using a carbohydrate-derived alkoxyallene as C3-building block. *Beilstein. J. Org. Chem.* **2013**, *9*, 2564–2569.
17. Dhand, V.; Chang, S.; Britton, R. Total Synthesis of the Cytotoxic Anhydrophytosphingosine Pachastrissamine (Jaspine B). *J. Org. Chem.* **2013**, *78*, 8208–8213. [CrossRef]
18. Enders, D.; Terteryan, V.; Paleček, J. Asymmetric Synthesis of Jaspine B (Pachastrissamine) via an Organocatalytic Aldol Reaction as Key Step. *Synthesis* **2008**, *14*, 2278–2282. [CrossRef]
19. Urano, H.; Enomoto, M.; Kuwahara, S. Enantioselective Syntheses of Pachastrissamine and Jaspine A via hydroxylactonization of a Chiral Epoxy Ester. *Biosci.* **2010**, *74*, 152–157.
20. Llaveria, J.; Díaz, Y.; Matheu, M.I.; Castillón, S. Eur. Enantioselective Synthesis of Jaspine B (Pachastrissamine) and Its C-2 and/or C-3 Epimers. *J. Org. Chem.* **2011**, 1514–1519.
21. Alnazer, H.; Castellan, T.; Salma, Y.; Génissom, Y.; Ballereau, S. Enantioselective Stereodivergent Synthesis of Jaspine B and 4-epi-Jaspine B from Axially Chiral Allenols. *Synlett* **2019**, *30*, 185–188.
22. Yakura, T.; Sato, S.; Yoshimoto, Y. Enantioselective Synthesis of Pachastrissamine (Jaspin B) Using Dirhodium(II)-Catalyzed C–H Amination and Asymmetric Dihydroxylation as Key Steps. *Chem. Pharm. Bull.* **2007**, *55*, 1284–1286. [CrossRef] [PubMed]

23. Abraham, E.; Candela-Lena, J.; Davies, S.G.; Georgiou, M.; Nicholson, R.L.; Roberts, P.M.; Russell, A.J.; Sánchez-Fernández, E.M.; Smith, A.D.; Thomson, J.E. Asymmetric synthesis of N,O,O,O-tetra-acetyl D-*lyxo*-phytosphingosine, jaspine B (pachastrissamine) and its C(2)-epimer. *Tetrahedron Asymmetry* **2007**, *18*, 2510–2513. [CrossRef]
24. Abraham, E.; Brock, E.A.; Candela-Lena, J.; Davies, S.G.; Georgiou, M.; Nicholson, R.L.; Perkins, J.H.; Roberts, P.M.; Russell, A.J.; Sánchez-Fernández, E.M.; et al. Asymmetric synthesis of N,O,O,O-tetra-acetyl D-*lyxo*-phytosphingosine, jaspine B (pachastrissamine), 2-*epi*-jaspine B, and deoxoprosophyllinevialithium amide conjugate addition. *Org. Biomol. Chem.* **2008**, *6*, 1665–1673.
25. Inuki, S.; Yoshimitsu, Y.; Oishi, S.; Fujii, N.; Ohno, H. Ring-Construction/Stereoselective Functionalization Cascade: Total Synthesis of Pachastrissamine (Jaspine B) through Palladium-Catalyzed Bis-cyclization of Bromoallenes. *Org. Lett.* **2009**, *11*, 4478–4481. [CrossRef] [PubMed]
26. Inuki, S.; Yoshimitsu, Y.; Oishi, S.; Fujii, N.; Ohno, H. Ring-Construction/Stereoselective Functionalization Cascade: Total Synthesis of Pachastrissamine (Jaspine B) through Palladium-Catalyzed Bis-cyclization of Propargyl Chlorides and Carbonates. *J. Org. Chem.* **2010**, *75*, 3831–3842. [CrossRef]
27. Yoshimitsu, Y.; Inuki, S.; Oishi, S.; Fujii, N.; Ohno, H. Stereoselective Divergent Synthesis of Four Diastereomers of Pachastrissamine(Jaspine B). *J. Org. Chem.* **2010**, *75*, 3843–3846. [CrossRef]
28. Passiniemi, M.; Koskinen, A.M.P. Asymmetric synthesis of Pachastrissamine (Jaspine B) and its diastereomers via η3-allylpalladium intermediates. *Org. Biomol. Chem.* **2011**, *9*, 1774–1783. [CrossRef] [PubMed]
29. Yoshimitsu, Y.; Miyagaki, J.; Oishi, S.; Fujii, N.; Ohno, H. Synthesis of pachastrissamine (jaspine B) and its derivatives by the late-stage introduction of the C-2 alkyl side-chains using olefin crossmetathesis. *Tetrahedron* **2013**, *69*, 4211–4220. [CrossRef]
30. Jana, A.K.; Panda, G. Stereoselective synthesis of Jaspine B and its C2 epimer from Garner aldehyde. *RSC Adv.* **2013**, *3*, 16795–16801. [CrossRef]
31. Yoshimitsu, Y.; Oishi, S.; Miyagaki, J.; Inuki, S.; Ohno, H.; Fujii, N. Pachastrissamine (jaspine B) and its stereoisomers inhibit sphingosine kinases and atypical protein kinase C. *Bioorg. Med. Chem.* **2011**, *19*, 5402–5408. [CrossRef] [PubMed]
32. Bae, H.; Jeon, H.; Baek, D.J.; Lee, D.; Kim, S. Stereochemically Reliable Syntheses of Pachastrissamine and Its 2-epi-Congener via Oxazolidinone Precursors from an Established Starting Material N-tert-Butoxycarbonyl-Protected Phytosphingosine. *Synthesis* **2012**, *44*, 3609–3612.
33. Vichare, P.; Chattopadhyay, A. Nitrolaldol reaction of (R)-2,3-cyclohexylideneglyceraldehyde: A simple and stereoselective synthesis of the cytotoxic Pachastrissamine (Jaspine B). *Tetrahedron Asymmetry* **2010**, *21*, 1983–1987. [CrossRef]
34. Ichikawa, Y.; Matsunaga, K.; Masuda, T.; Kotsuki, H.; Nakano, K. Stereocontrolled synthesis of cytotoxic anhydrosphingosine pachastrissamine by using [3.3] sigmatropic rearrangement of allyl cyanate. *Tetrahedron* **2008**, *64*, 11313–11318. [CrossRef]
35. Bhaket, P.; Morris, K.; Stauffer, C.S.; Datta, A. Total Synthesis of Cytotoxic Anhydrophytosphingosine Pachastrissamine (Jaspine B). *Org. Lett.* **2005**, *7*, 875–876. [CrossRef]
36. Lee, H.-J.; Lim, C.; Hwang, S.; Jeong, B.-S.; Kim, S. Silver-Mediated exo-Selective Tandem Desilylative Bromination/Oxycyclization of Silyl-Protected Alkynes: Synthesis of 2-Bromomethylene-Tetrahydrofuran. *Chem. Asian J.* **2011**, *6*, 1943–1947. [CrossRef]
37. Prasad, K.R.; Chandrakumar, A. Stereoselective Synthesis of Cytotoxic Anhydrophytosphingosine Pachastrissamine [Jaspine B]. *J. Org. Chem.* **2007**, *72*, 6312–6315. [CrossRef]
38. Lee, T.; Lee, S.; Kwak, Y.S.; Kim, D.; Kim, S. Synthesis of Pachastrissamine from Phytosphingosine: A Comparison of Cyclic Sulfate vs an Epoxide Intermediate in Cyclization. *Org. Lett.* **2007**, *9*, 429–432. [CrossRef] [PubMed]
39. Van der Berg, R.J.B.H.N.; Boltje, T.J.; Verhagen, C.P.; Litjens, E.J.N.; Van der Marel, G.A.; Overkleeft, H.S. An Efficient Synthesis of the Natural Tetrahydrofuran Pachastrissamine Starting from D-*ribo*-Phytosphingosine. *J. Org. Chem.* **2006**, *71*, 836–839. [CrossRef]
40. Fujiwara, T.; Liu, B.; Niu, W.; Hashimoto, K.; Nambu, H.; Yakura, T. Practical Synthesis of Pachastrissamine (Jaspine B), 2-*epi*-Pachastrissamine, and the 2-*epi*-Pyrrolidine Analogue. *Chem. Pharm. Bull.* **2016**, *64*, 179–188. [CrossRef]
41. Du, Y.; Liu, J.; Linhardt, R.J. Stereoselective Synthesis of Cytotoxic Anhydrophytosphingosine Pachastrissamine (Jaspine B) from D-Xylose. *J. Org. Chem.* **2006**, *71*, 1251–1253. [CrossRef] [PubMed]
42. Reddy, L.V.R.; Reddy, P.V.; Shaw, A.K. An expedient route for the practical synthesis of pachastrissamine (jaspine B) starting from 3,4,6-tri-O-benzyl-D-galactal. *Tetrahedron Asymmetry* **2007**, *18*, 542–546. [CrossRef]
43. Ribes, C.; Falomir, E.; Carda, M.; Marco, J.A. Stereoselective synthesis of pachastrissamine (jaspine B). *Tetrahedron* **2006**, *62*, 5421–5425. [CrossRef]
44. Ramana, C.V.; Giri, A.G.; Suryawanshi, S.B.; Gonnade, R.G. Total synthesis of pachastrissamine (jaspine B) enantiomers from d-glucose. *Tetrahedron Lett.* **2007**, *48*, 265–268. [CrossRef]
45. Rao, G.S.; Rao, B.V. A common strategy for the stereoselective synthesis of anhydrophytosphingosine (jaspine B) and N,O,O,O-tetra-acetyl D-*lyxo*-phytosphingosine. *Tetrahedron Lett.* **2011**, *52*, 6076–6079.
46. Kwon, Y.; Song, J.; Bae, H.; Kim, W.J.; Lee, J.Y.; Han, G.H.; Lee, S.K.; Kim, S. Synthesis and Biological Evaluation of Carbocyclic Analogues of Pachastrissamine. *Mar. Drugs* **2015**, *13*, 824–837. [CrossRef] [PubMed]
47. Lopatka, P.; Markovič, M.; Koóš, P.; Ley, S.L.; Gracza, T. Continuous Pd-Catalyzed Carbonylative Cyclization Using Iron Pentacarbonyl as a CO Source. *J. Org. Chem.* **2019**, *84*, 14394–14406. [CrossRef] [PubMed]
48. Markovič, M.; Koóš, P.; Gracza, T. A Short Asymmetric Synthesis of Sauropunols A–D. *Synthesis* **2017**, *49*, 2939–2942.
49. Markoviš, M.; Koóš, P.; Čarný, T.; Sokoliová, S.; Boháčiková, N.; Moncoľ, J.; Gracza, T. Total Synthesis, Configuration Assignment, and Cytotoxic Activity Evaluation of Protulactone A. *J. Nat. Prod.* **2017**, *80*, 1631–1638. [CrossRef]

50. Lopatka, P.; Koóš, P.; Markovič, M.; Gracza, T. Asymmetric Formal Synthesis of (+)-Pyrenolide D. *Synthesis* **2014**, *46*, 817–821.
51. Koóš, P.; Špánik, I.; Gracza, T. Asymmetric intramolecular Pd (II)-catalysed amidocarbonylation of unsaturated amino alcohols. *Tetrahedron Asymmetry* **2009**, *20*, 2720–2723. [CrossRef]
52. Kapitán, P.; Gracza, T. Stereocontrolled oxycarbonylation of 4-benzyloxyhepta-1,6-diene-3,5-diols promoted by chiral palladium (II) complexes. *Tetrahedron Asymmetry* **2008**, *19*, 38–44. [CrossRef]
53. McKillop, A.; Taylor, R.; Watson, R.; Lewis, N. An Improved Procedure for the Preparation of the Garner Aldehyde and Its Use for the Synthesis of N-Protected 1-Halo-2(R)-amino-3-butanes. *Synthesis* **1994**, *1*, 31–33. [CrossRef]
54. Ojima, I.; Vidal, S.E. Rhodium-Catalyzed Cyclohydrocarbonylation: Application to the Synthesis of (+)-Prosopinine and (−)-Deoxoprosophylline. *J. Org. Chem.* **1998**, *63*, 7999–8003. [CrossRef]
55. Ghosh, A.; Chattopadhyay, S.K. A diversity oriented synthesis of D-erythro-sphingosine and siblings. *Tetrahedron Asymmetry* **2017**, *28*, 1139–1143. [CrossRef]
56. Ghosal, P.; Shaw, A.K. An efficient total synthesis of the anticancer agent (+)-spisulosine (ES-285) from Garner's aldehyde. *Tetrahedron Lett.* **2010**, *51*, 4140–4142. [CrossRef]
57. Singh, P.; Panda, G. Intramolecular 5-*endo*-trig aminopalladation of β-hydroxy-γ-alkenylamine: Efficient route to apyrrolidine ring and its application for the synthesis of (-)-8,8a-di-*epi*-swainsonine. *RSC Adv.* **2014**, *4*, 2161–2166. [CrossRef]
58. Abraham, E.; Davies, S.G.; Roberts, P.M.; Russell, A.J.; Thomson, J.E. Jaspine B (pachastrissamine) and 2-epi-jaspine B:synthesis and structural assignment. *Tetrahedron Asymmetry* **2008**, *19*, 1027. [CrossRef]
59. Porta, R.; Benaglia, M.; Puglisi, A. Flow Chemistry: Recent Developments in the Synthesis of Pharmaceutical Products. *Org. Preocess Res. Dev.* **2016**, *1*, 2–25. [CrossRef]
60. Brandão, P.; Pineiro, M.; Pinho e Melo, T.M. Flow Chemistry: Towards A More Sustainable Heterocyclic Synthesis. *Eur. J. Org. Chem.* **2019**, *43*, 7188–7217. [CrossRef]
61. Levesque, F.; Seeberger, F.H. Continuous-Flow Synthesis of the Anti-Malaria Drug Artemisinin. *Angew. Chem. Int. Ed.* **2012**, *51*, 1706–1709. [CrossRef]
62. Lau, S.-H.; Galván, A.; Merchant, R.R.; Battilocchio, C.; Souto, J.A.; Berry, M.B.; Ley, S.V. Machines vs Malaria: A Flow-Based Preparation of the Drug Candidate OZ439. *Org. Lett.* **2015**, *17*, 3218–3221. [CrossRef] [PubMed]
63. LaPorte, T.L.; Spangler, L.; Hamedi, M.; Lobben, P.; Chan, S.H.; Mushlehiddinoglu, J.; Wang, S.S.Y. Development of a Continuous Plug Flow Process for Preparation of a Key Intermediate for Brivanib Alaninate. *Org. Process Res. Dev.* **2014**, *18*, 1492–1502. [CrossRef]
64. Baxendale, I.R.; Griffiths-Jones, C.M.; Ley, S.V.; Tranmer, G.K. Preparation of the Neolignan Natural Product Grossamide by a Continuous-Flow Process. *Synlett* **2006**, *3*, 427–430. [CrossRef]
65. Correia, C.A.; Gilmore, K.; McQuade, D.T.; Seeberger, P.H. A Cocise Flow Synthesis of Efavirenz. *Angew. Chem. Int. Ed.* **2015**, *54*, 4945–4948. [CrossRef] [PubMed]
66. Tsubogo, T.; Oyamada, H.; Kobayashi, S. Multistep continuous-flow synthesis of (R)- and (S)-rolipram using heterogeneous catalysts. *Nature* **2015**, *520*, 329–332. [CrossRef]
67. Hartwig, J.; Ceylan, S.; Kupracz, L.; Coutable, L.; Kirschning, A. Heating under High-Frequency Inductive Conditions: Application to the Continuous Synthesis of the Neurolepticum Olanzapine (Zyprexa). *Angew. Chem. Int. Ed.* **2013**, *52*, 9813–9817. [CrossRef] [PubMed]
68. Babjak, M.; Markovič, M.; Kandríková, B.; Gracza, T. Homogeneous Cyclocarbonylation of Alkenols with Iron Pentacarbonyl. *Synthesis* **2014**, *46*, 809–816. [CrossRef]

Review

Evolution of Pauson-Khand Reaction: Strategic Applications in Total Syntheses of Architecturally Complex Natural Products (2016–2020)

Sijia Chen [1,†], Chongguo Jiang [1,†], Nan Zheng [1], Zhen Yang [1,2,3,*] and Lili Shi [1,*]

1. State Key Laboratory of Chemical Oncogenomics and Key Laboratory of Chemical Genomics, Peking University Shenzhen Graduate School, Shenzhen 518055, China; chensijia941216@163.com (S.C.); jiangcg@pku.edu.cn (C.J.); zhengnan123@pku.edu.cn (N.Z.)
2. Key Laboratory of Bioorganic Chemistry and Molecular Engineering of Ministry of Education and Beijing National Laboratory for Molecular Science (BNLMS), and Peking-Tsinghua Center for Life Sciences, Peking University, Beijing 100871, China
3. Shenzhen Bay Laboratory, Shenzhen 518055, China
* Correspondence: zyang@pku.edu.cn (Z.Y.); shill@pkusz.edu.cn (L.S.)
† Sijia Chen and Chongguo Jiang have contributed equally to this work.

Received: 23 September 2020; Accepted: 14 October 2020; Published: 16 October 2020

Abstract: Metal-mediated cyclizations are important transformations in a natural product total synthesis. The Pauson-Khand reaction, particularly powerful for establishing cyclopentenone-containing structures, is distinguished as one of the most attractive annulation processes routinely employed in synthesis campaigns. This review covers Co, Rh, and Pd catalyzed Pauson-Khand reaction and summarizes its strategic applications in total syntheses of structurally complex natural products in the last five years. Additionally, the hetero-Pauson-Khand reaction in the synthesis of heterocycles will also be discussed. Focusing on the panorama of organic synthesis, this review highlights the strategically developed Pauson-Khand reaction in fulfilling total synthetic tasks and its synthetic attractiveness is aimed to be illustrated.

Keywords: metal-mediated reactions; Pauson-Khand reaction; cyclopentenones; natural products total syntheses

1. Introduction

The metal-mediated reaction plays an important role in constructing complex organic molecules [1–3]. The Pauson-Khand reaction (PKR), an effective set of annulation protocol defined in 1973 [4] for the construction of cyclopentenone-containing moieties, stands as a promising method to permit efficient cyclic frameworks. Its efficient and atom-economic elaboration to substituted cyclopentenones renders this process highly prized in the construction of architecturally complex natural products. Since reported more than 40 years ago [5–12], it has been developed with different metal catalytic systems, including Co [13–17], Rh [18–25], Ru [26–30], Ti [31–34], Ir [35–37], Ni [38], Mo [39,40], Fe [41]; and other metals could promote the PKR to build the heterocycle frameworks [42–44]. By identifying reactivity patterns for diverse PKR precursors in the prominent synthetic application, we aim to elevate this powerful reaction to a method of choice in the synthetic designation of complex biologically active entities.

1.1. Classic PK Reaction Catalyzed by Co

In 1973, I.U. Khand and P.L. Pauson found that the generation of enyne/$Co_2(CO)_6$ complex with olefin as substrates could lead to the formation of cyclopentenone. Moving forward, P.L. Pauson explored the substrate scope and limitations of this reaction [45]. Although the specific mechanism of PKR involving Co$_2$(CO)$_8$ is still uncertain, the mechanism proposed by Magnus [46–48] and Schore [49] is widely recognized based on the reaction results of regioselectivity and stereoselectivity (Scheme 1). The rate-determining step is alkene coordination with the cobalt and then insertion into cobalt–carbon bond to form the cobaltacycle, accounting for the regiochemical and stereochemical outcomes.

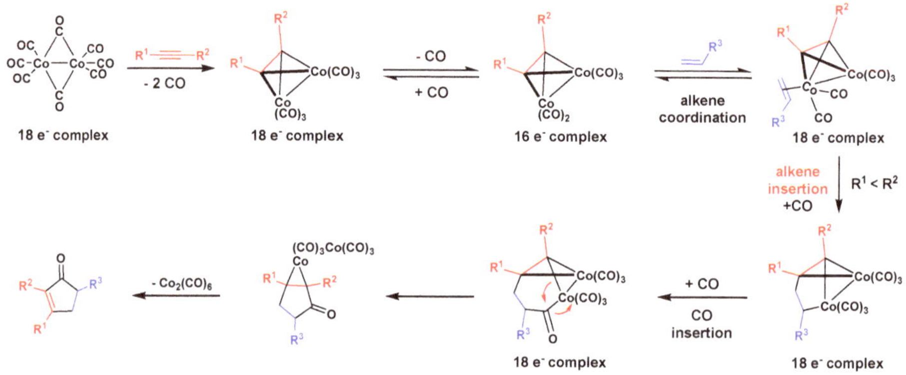

Scheme 1. Generally accepted mechanism of Pauson-Khand reaction with Co$_2$(CO)$_8$.

The regioselectivity of PKR is influenced by both steric and electronic effects (Scheme 2). For electrically neutral substrates, the insertion of olefins to enyne/Co$_2$(CO)$_6$ complex correlates with steric hindrance. The regioselectivity also has been demonstrated to be related to the electronegativity of alkynyl groups [50–52]. Under most circumstances, the electron-withdrawing group will be installed at the β position of cyclopentenone. It is noteworthy that the frontier molecular orbital (FMO) theory could be used to analyze the influence of olefins in PKR [53,54]. Moreover, subordinate interaction and the guiding group can affect the regioselectivity [55–57]. For allene-involved intramolecular PKR, a 5,7-bicyclic product is more inclined to be formed [58,59].

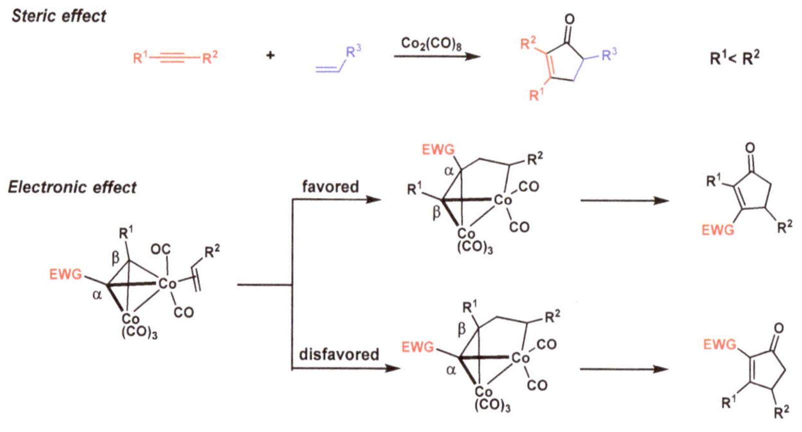

Scheme 2. Regioselectivity study of Pauson-Khand reaction.

As for the diastereoselectivity of intramolecular PKR, both substrate conformation (especially the allyl chiral center) and electronic effect are relevant parameters (Scheme 3). Krafft reported their reaction with electron-deficient alkynes, and the PKR product could be obtained with a high *dr* value when norbornene was involved as an olefin substrate [51,60].

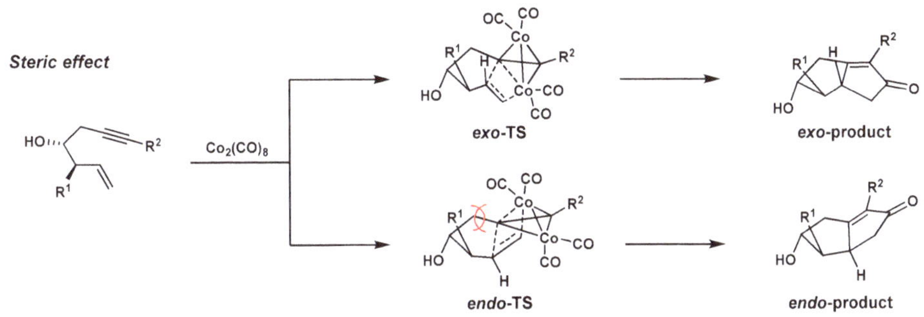

Scheme 3. Diastereoselectivity study of intramolecular Pauson-Khand reaction.

Most of the Co-catalyzed PKR conditions require a relatively high temperature and long reaction time. To accelerate the reaction rate, Smit and Caple's group found that PKR could be promoted in a stepwise manner [61]. N-methylmorpholine oxide (NMO), acting as an additive, was reported to improve the reaction rate through oxidizing CO into CO_2 on the enyne/$Co_2(CO)_6$ complex [62], forcing the cobalt to release a vacant orbital which can be coordinated with olefins [63]. Recently, the Dionicio Martinez-Solorio group demonstrated the value of 4-FBnSMe as a new, efficient, and recoverable/reusable thioether promoter in PKR by modulating the Lewis basicity of thioether to influence the rate of alkene insertion [64].

To circumvent the use of stoichiometric catalysts, some Lewis bases were discovered to achieve the catalytic version of PKR, such as phosphine ligand [65], tetramethyl thiourea [66], phosphane sulfide [14], and primary amines [67]. In 2005, the Milet and Gimbert groups converted to the density functional theory (DFT) and calculated the energy change of the PKR process with Lewis base [68]. The results indicate that the enyne/$Co_2(CO)_6$-alkene insertion is a reversible process, but the Lewis base coordination could reduce the energy and therefore make the olefin insertion process irreversible.

1.2. PK Reaction Catalyzed by Rh and Pd

The first example of $[RhCl(CO)_2]_2$ catalyzed PKR was reported by Narasaka et al. in 1998 [18]. In their studies, the use of toluene as a reaction media reduced the loading of Rh catalysts and a good reaction reactivity was achieved with electron-deficient alkynes [69]. Moreover, under a low partial pressure of CO, it can effectively speed up the reaction and decrease the reaction temperature. Jeong et al. reported the first case of rhodium-catalyzed asymmetric PKR in the presence of 2,2'-*bis*(di-*p*-tolylphosphino)-1,1'-binaphthyl(BINAP) and AgOTf [21]. Consiglio's group used the molecular sieve to adsorb CO, greatly reducing the reaction temperature and accelerating the rate [70]. They accomplished the asymmetric PKR at 0 °C with a 99% *ee* value. In the course of Wender and his co-workers' studies on the rhodium(I)-catalyzed intra- and intermolecular dienyl [2 + 2 + 1] PKR, they observed that when a diene was used in place of an alkene the reaction rate was significantly accelerated [71,72].

As the Rh-catalyzed PKR has several advantages, it has attracted the attention of many research groups to report their work in this area. Typical PKR requires the utilization of highly toxic CO gas. An important breakthrough was made by the Morimoto and Shibata groups, respectively by introducing metal carbonyl compounds as a masked CO source through transition metal decarbonylation to in situ generate CO in PKR [73]. Moreover, Chung's group developed the use of a highly beneficial cinnamyl-alcohol as a CO source in the presence of the Rh catalyst to obtain corresponding hetero-Pauson-Khand (hPK) products in an inexpensive, safe, and environmentally friendly manner [74]. The catalytic dehydrogenation of cinnamyl alcohol could produce cinnamaldehyde, followed by Wilkinson decarbonylation and carbonylation constructed the desired cyclic product. Benzyl formate [75] has also been exploited as a non-gas CO surrogate. In 2019, they further demonstrated the utilization of the formic acid as a CO source in the formation of various bicyclic cyclopentenones. In their protocol, formic acid was employed as a bridging molecule for the conversion of CO_2 to CO, which represented an indirect approach for the chemical valorization of CO_2 in the construction of valuable heterocycles [76] (Scheme 4).

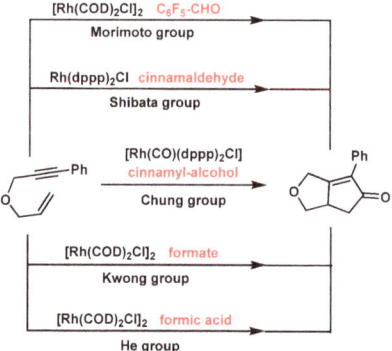

Scheme 4. Non-gas CO surrogates in the Pauson-Khand reaction for heterocycle's formation.

The theoretical analysis of Rh-catalyzed PKR diastereoselectivity was demonstrated by Baik's group [25]. They revealed that two possible mechanistic scenarios and the optimum selectivity could be attributed to a five-coordinate organorhodium complex. The larger energy gap between the diastereomers and the Rh meta-cyclization trend to occur at the *cis*-position site dominated the diastereoselectivity. Based on the high efficiency, reliability and excellent diastereoselectivity of Rh-catalyzed PKR, its extraordinary impact on the synthetic campaign as a key step has been recognized [77].

Few metals could be applied in the catalytic PKR (Co, Ti, Rh, Ir, Ru) as most of them are air and moisture sensitive, and as such, it accounted for some limitations in synthetic applications. A series of thiourea and Pd-catalyzed reactions were developed by Yang's group [78–81] (Scheme 5). $PdCl_2$ coordinated to a thiourea ligand could catalyze an intramolecular PKR under mild conditions [81,82], and some interesting features were observed in this novel step. It could be catalyzed by $PdCl_2$ alone with a low yield, whereas using thiourea, especially tetramethyl thiourea (TMTU), as a reaction additive could greatly increase the yield; the Lewis acids addition such as LiCl can increase both the reaction rate and yield. Based on this observed phenomenon, further DFT calculation and mechanism investigation were carried out [83]. According to the coordination mode of the transition state, the TMTU ligand and substrate are lying both on the same side of the Pd catalyst thus the trans-diastereomer in substituted cases outperformed its diastereoisomer. It is speculated that changes in thiourea ligands may affect the diastereoselectivity of PKR through steric effect and π–π interaction, etc.

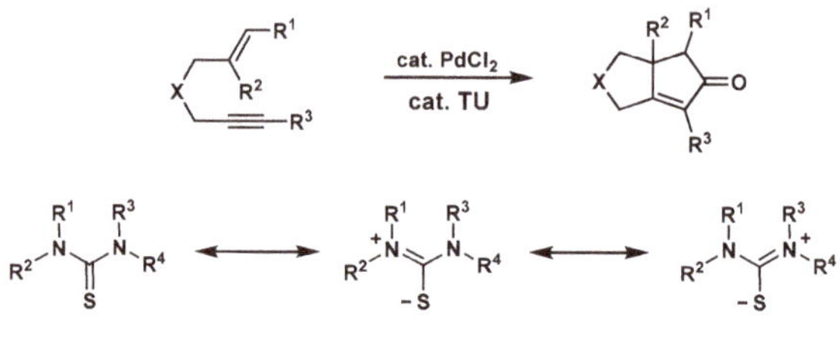

Scheme 5. Tetramethyl thiourea (TMTU) and Pd-catalyzed Pauson-Khand reaction.

1.3. Hetero-Pauson-Khand Reaction

The hetero-Pauson-Khand reaction has been harnessed as an effective tactic in the concise construction of functionalized polycyclic butenolides and α, β-unsaturated lactams (Scheme 6). In 1996, Crowe et al. reported the direct synthesis of bicyclic γ-butyrolactones via tandem reductive cyclization-carbonylation of tethered enals and enones [84,85]. In the same year, Buchwald et al. presented the heteroatom variant of the intramolecular PKR catalyzed by Cp$_2$Ti(PMe$_3$)$_2$, in which the alkene could be replaced with a carbonyl for the diastereoselective synthesis of γ-butyrolactones or a fused butenolide, respectively [86,87]. Later on, chemists devoted themselves in the development of hetero-Pauson-Khand reaction, including Murai [28,88], Carretero [89], Saito [90], and Snapper [91]. However, the application of hPK in a natural product total synthetic work is relatively rare and therefore is underexplored in the synthetic version [92–96].

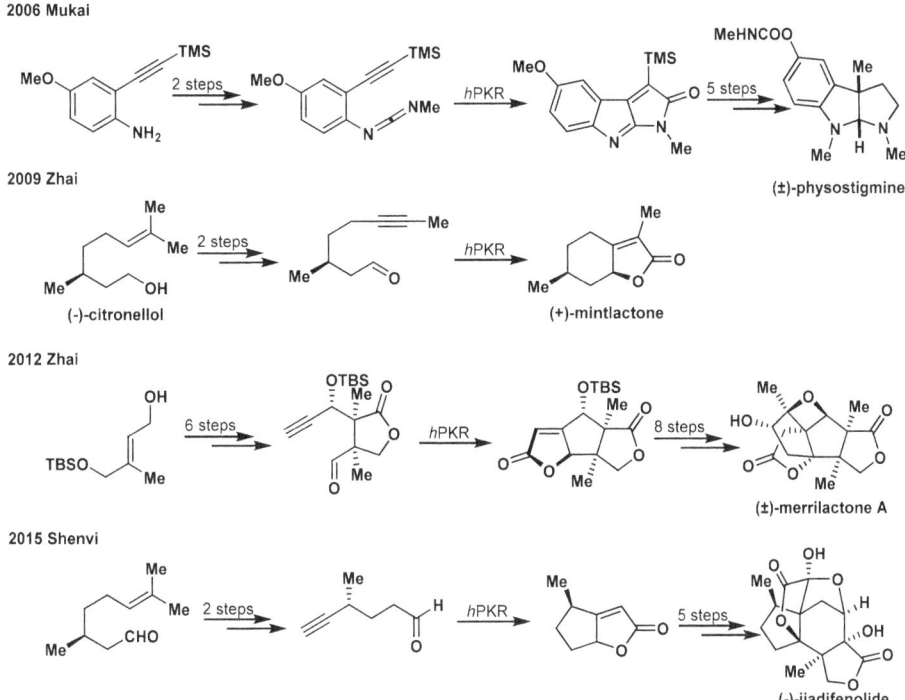

Scheme 6. Hetero-Pauson-Khand reaction in natural products total syntheses.

1.4. Summary

Cobalt, rhodium, and palladium were involved in PK reactions represented in different advantageous patterns, among which the outstanding superiorities are as follows: a. Cobalt-catalyzed PKRs can overcome the high tension and construct an all-carbon quaternary chiral center [7,97–99]; b. rhodium-catalyzed PKRs normally exhibit excellent diastereoselectivity and are attractive in building a variety of ring structures; c. palladium-catalyzed PKRs could lead to the opposite stereoselectivity compared with others and are more operable due to the stability of Palladium species. Co/Rh-catalyzed PKRs are already widely applied in natural products total syntheses, in contrast, restriction existed in Pd-catalyzed PKRs and most of the work is still under methodological study.

The stereoselective formation of quaternary chiral centers is challenging in the construction of the cyclic system. PKR is an effective method for generating 5,5-bicyclic ring systems and has already been studied comprehensively. In 1984, Schore's group reported the first case of PKR to construct a 5,5,5-tricyclic skeleton containing an all-carbon quaternary chiral center [100]. Numerous research groups reported their studies and applications in natural products total syntheses. Joseph M. Fox et al. applied a thiourea-facilitated PKR in establishing the quaternary center and built a 5,5,3-tricyclic framework, and then completed the enantioselective total synthesis of (−)-pentalenene [101]. In the past few years, many chemists have made their efforts to broaden the application of the intramolecular PKR in natural products total syntheses, with some reviews already published [5–12]. In this mini-review, a perspective on the development of strategic Pauson-Khand reaction within natural products total syntheses portfolio over the past five years is presented by the categories of the constructed bicyclic ring systems (5,5/5,6/5,7- and macrocycles), with the aim to provide an updated overview of its tremendous power and versatility.

2. Recent Pauson-Khand Reaction Applications in Natural Products Total Syntheses

PKR proved to be a powerful strategy in natural products syntheses, particularly in those containing fused five-membered rings. The tethered length plays an important role in the efficiency and viability of all intramoleculars.

Pauson-Khand-like reactions [102], and the substrates with tethers that result in the formation of a five-membered ring are most effective in a great variety of intramolecular reactions [103]. Collections of 5,5/5,6/5,7-bicyclic ring systems or even macrocycles could be accessed depending on substrate identity as shown in this review.

2.1. Construction of 5,5-Bicyclic Ring Systems

Ryanodol is a bioactive and complex poly-alcohol containing natural product which is a potent modulator of the calcium release channel [104,105]. In 2016, Reisman's group reported a highly efficient way to rapidly build the carbon framework of ryanodol through intramolecular PKR which was promoted by the rigidity of the bicyclic conformation [106]. Starting from S-pulegone, the PK precursor **1** could be achieved after seven steps of transformation. In their promising reaction protocol, submitting **1** with 1 mol% [RhCl(CO)$_2$]$_2$ under an atmosphere of CO afforded enone **2** in an 85% yield as a single diastereomer. More impressively, the efficient protocol could be performed on the multi-gram scale and provided a 5.7 g of PK product (Scheme 7).

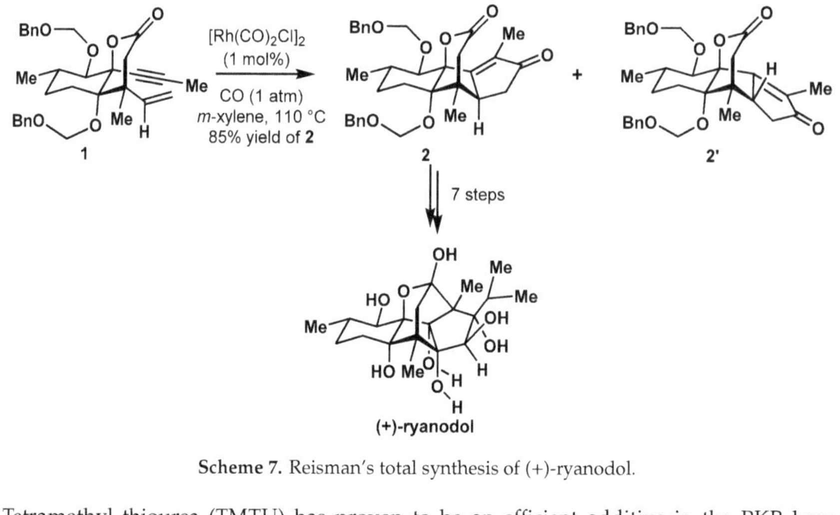

Scheme 7. Reisman's total synthesis of (+)-ryanodol.

Tetramethyl thiourea (TMTU) has proven to be an efficient additive in the PKR based on Yang's previous investigations [66]. In 2017, they developed a Co–TMTU catalyzed PKR and 6π electrocyclization tandem reactions to construct the highly strained core skeleton of presilphiperfolanols and related natural products [107,108]. Treatment of **3** with a catalytic amount of [Co$_2$(CO)$_8$] (0.2 equiv.) and TMTU (1.2 equiv.) in benzene resulted in the rapid construction of the tricyclic scaffold **5** with great regio- and stereochemical control in a 94% yield through one single operation. Most recently, they applied this PKR model to the synthesis of 4-desmethyl-rippertenol and 7-epi-rippertenol [109] (Scheme 8).

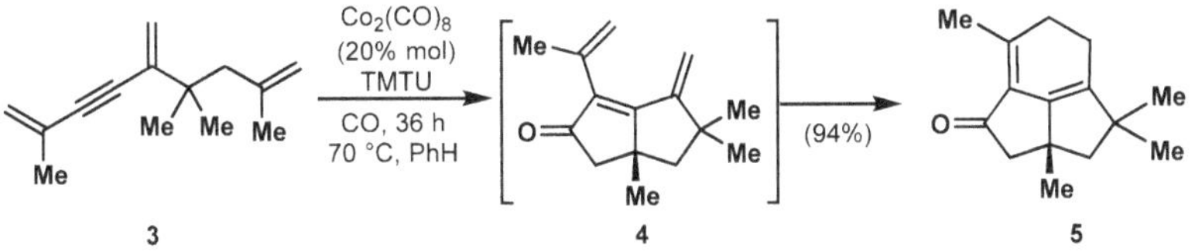

Scheme 8. Yang's concise synthesis of presilphiperfolane core.

In the total synthesis of the potent antibiotic compounds (−)-crinipellin A and (−)-crinipellin B reported by Yang et al. [97], the fully functionalized tetraquinane core was achieved by a novel thiourea/palladium-catalyzed PKR. They implemented two PKRs in their synthetic strategy with the first being a conversion of **6** into **7** with a 40% yield and 98% ee after crystallization. As generally proved, the electron-deficient alkyne is not a perfect ligand for Co$_2$(CO)$_8$, gradual warming is essential for constructing the desired enyne/Co$_2$(CO)$_6$ complex. The other PKR allows the concise formation of the tetraquinane **10**. Treatment of **8** in the presence of NaHCO$_3$ as the base provided the desired tetraquinane core **10** in a 61% yield, with the undesired isomer suppressed to a 16% yield (Scheme 9).

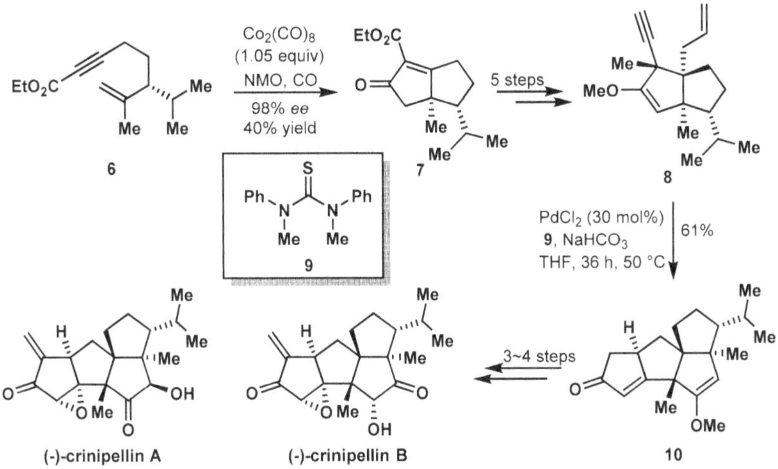

Scheme 9. Yang's total synthesis of crinipellins.

In 2018, Yang's group described a stereoselective construction of the CDEFGH ring system of lancifodilactone G acetate and a 28-step asymmetric total synthesis [110,111]. They performed an intramolecular PKR for the construction of the sterically congested F ring. In their model study, the authors observed that the butynoic ester was effective for the regio- and stereoselectivity in constructing the cyclopentenone ring system bearing two chiral centers. The developed well-orchestrated PKR facilitated the stereoselective synthesis of **14** from enyne **13** (Scheme 10).

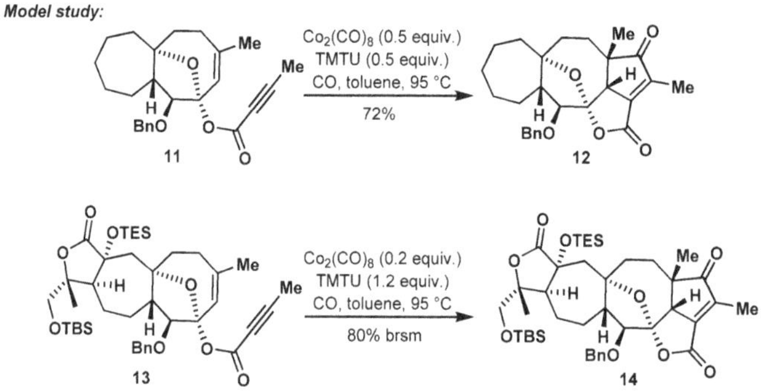

Scheme 10. Yang's asymmetric total synthesis of lancifodilactone G acetate.

Li et al. reported the first total synthetic work of hybridaphniphylline B featuring a late-stage intermolecular Diels–Alder reaction [112]. They implemented a PKR and C=C bond migration strategy to achieve the key intermediate **17**. Through the investigation of Pauson-Khand conditions, it is determined that MeCN is an effective accelerator to transform the alkyne dicobalt complex into the desired product, which was depicted the same as Pauson's work [113]. Under this condition, the two PKR products **16** and **16'** were constructed in a 73% yield with the ratio of ca. 2.4:1. Then, submitting the mixture to K_2CO_3/TFE realized the C=C bond migration and gave the more substituted enone **17** (Scheme 11).

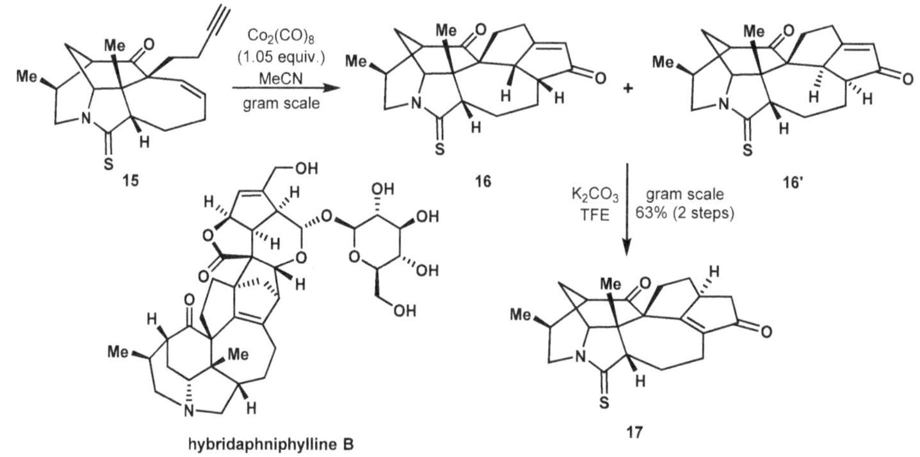

Scheme 11. Li's total synthesis of hybridaphniphylline B.

Liang et al. described a concise total synthesis towards (−)-indoxamycins A and B, a novel class of polyketide natural products, which contain a highly congested cage-like carbon skeleton featuring six contiguous chiral centers [114]. The key step for rapidly constructing the framework bearing a quaternary carbon was an intramolecular PKR. Enyne **18** was converted into the 5,5,6-tricyclic compound **19** smoothly with a 74% yield, which could be further transformed into the target natural products (Scheme 12).

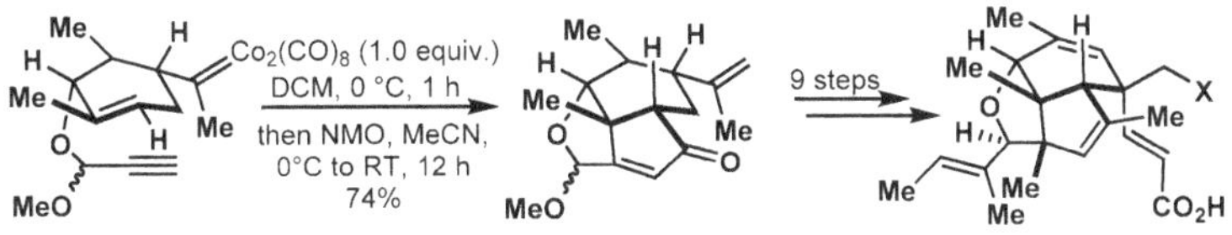

Scheme 12. Liang's total synthesis of (−)-indoxamycins A and B.

Guaianolide sesquiterpenes represent a particularly prolific class of terpene natural products, which have attracted biological and chemical communities for decades given their extensive documented therapeutic properties and fascinating chemical structures. Recently, the cobalt-mediated intramolecular PKR was applied in the total synthesis of sinodielide A and ent-absinthin by Mainone et al. [115]. Ester 20, converted from (−)-linalool via deprotonation and a subsequent reaction with the mixed anhydride of 2-butynoic acid, underwent a smooth PKR reaction using $Co_2(CO)_8$ and resulted in strained bicyclic lactone 21 (65% yield, 5:2 d.r.), which enabled concise and collective total syntheses of guaianolide sesquiterpenes (Scheme 13).

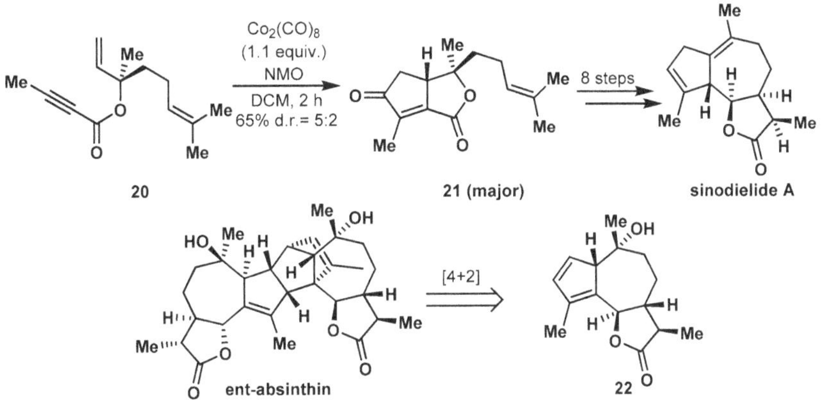

Scheme 13. Maimone's allylative approaches to the synthesis of complex guaianolide sesquiterpenes.

Yang et al. recently described the first asymmetric total synthesis of (−)-spirochensilide A featuring a tungsten-mediated cyclopropene-based PKR to install the quaternary chiral center [116]. Initially, they attempted various conditions to construct the cyclopentenone motif in 24 but all proved in vain presumably due to the low reactivity of enyne 23 and its steric rigidity. Recognizing the inherent of a chloride, they employed it as an σ electron-withdrawing group to promote polarization and reduce the activation barrier, with the idea in hand they prepared chloroenyne 25. However, reaction conditions screening only resulted in the undesired ring-closing compounds 26 and 27, respectively, which was generated by an Rh-catalyzed carbonylative C-H insertion and a double bond isomerization followed by a PKR. Then, they considered taking advantage of cyclopropene's inherent strain and altered the pathway to construct enyne 28. After an investigation of many conditions, the $W(CO)_3(MeCN)_3$, $Ni(COD)_2$/bipy, and $Mo(CO)_3(DMF)_3$-catalyzed PKR could lead to the formation of the desired 29a, which could be further transformed into (−)-spirochensilide A (Scheme 14).

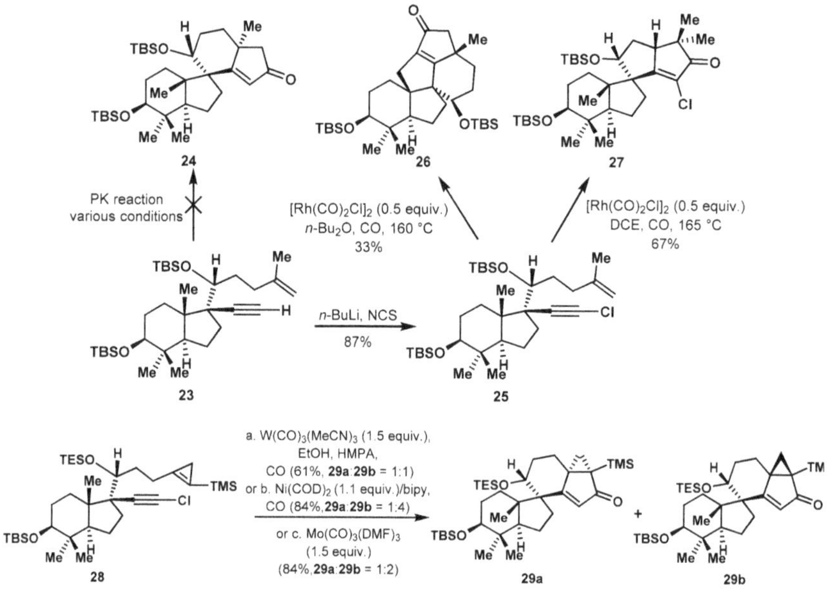

Scheme 14. Yang's asymmetric total synthesis of (−)-spirochensilide A.

Very recently, Yang and Snyder's group both reported their total synthesis towards the challenging target (+)-waihoensene [98,99], which contains four contiguous quaternary carbon centers. In their strategies, they all involved a diastereoselective Conia-ene type reaction and an intramolecular PKR. The polycyclic skeleton of Waihoensene was achieved by the $Co_2(CO)_8$-mediated PKR under CO atmosphere (Scheme 15).

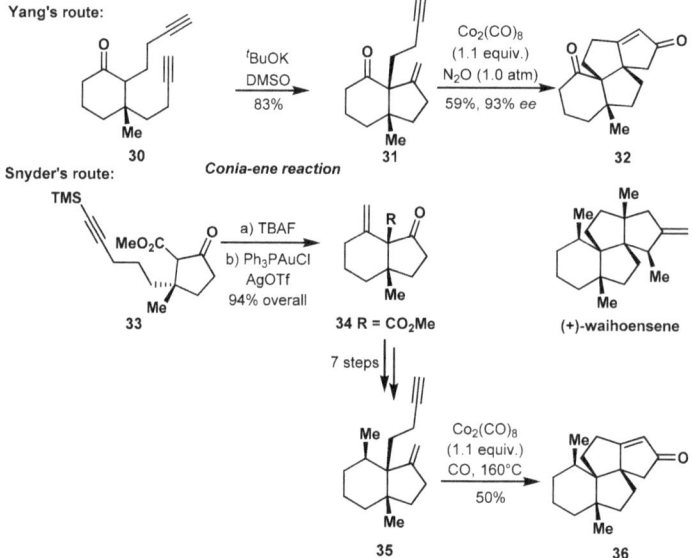

Scheme 15. Yang's and Snyder's total synthesis of (+)-waihoensene.

(−)-Conidiogenone B, (−)-conidiogenone, and (−)-conidiogenol feature a highly strained 6/5/5/5 tetracyclic core and 6-8 consecutive stereocenters. The concise total syntheses have been accomplished by Zhai et al. [117]. The key linear triquinane **38** was constructed as a single diastereomer in a 71% yield via a tandem Nicholas and amine-N-oxide-promoted PKR from **37** with the borane-methyl sulfide complex as the hydride source (Scheme 16).

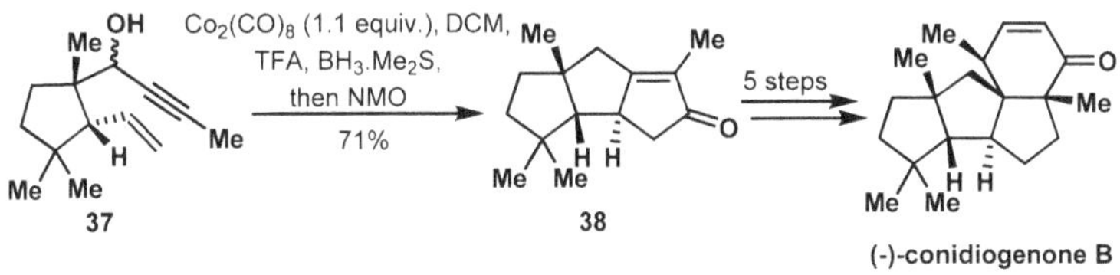

Scheme 16. Zhai's total synthesis of (−)-conidiogenone B.

2.2. Construction of 5,6-Bicyclic Ring Systems

Clark et al. elucidated an efficient 12-step synthesis of the marine alkaloid (−)-nakadomarin A [118], which contains a unique hexacyclic structure featuring fused 5-, 6-, 8-, and 15- membered rings and exhibits cytotoxicity against murine lymphoma L1210 cells, antimicrobial and inhibitory activity against cyclin-dependent kinase 4. The fused bicyclic enone **40** was constructed in a good yield and with an excellent *ee* value using the asymmetric cobalt-catalyzed PKR, which was developed by Hiroi et al. earlier before (Scheme 17).

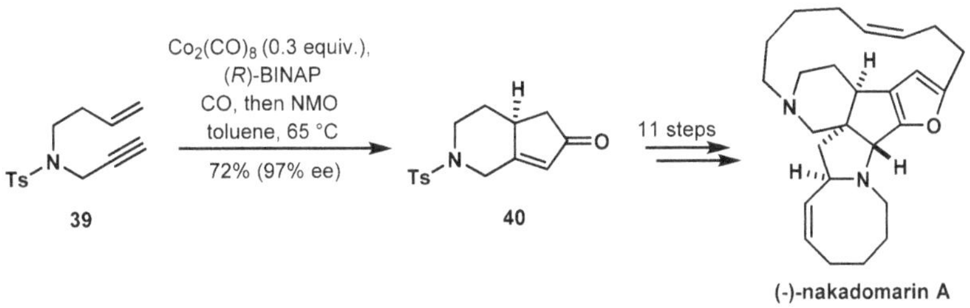

Scheme 17. Clark's total synthesis of (−)-nakadomarin A.

In the Hao et al. studies toward the 10 step-synthesis of a novel limonoid perforanoid A [119], they investigated Rh-catalyzed intramolecular PKR to build the cyclopentenone ring. Under their optimum conditions, treatment of **42** in toluene for 3 h at a reaction concentration of 8 mM with [Rh(CO)$_2$Cl]$_2$ (7 mol%) as the catalyst gave **43** in 85% as a single isomer (Scheme 18).

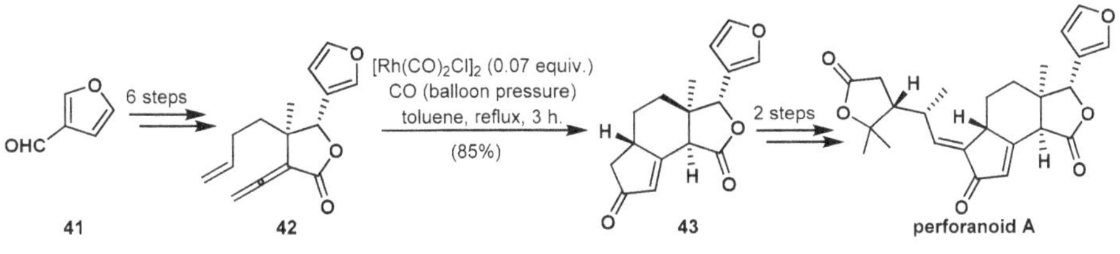

Scheme 18. Hao's asymmetric total synthesis of perforanoid A.

Zard, Takayama, and Mukai groups have explored the diastereoselective study of intramolecular PKR in the context of Lycopodium alkaloids syntheses [120–122]. Based on their previous study, Trauner et al. used a similar strategy to synthesize enone **46** with the desired stereoselectivity, which was proposed through a chair-like conformation of intermediate **45** ensuing the bicycle [4.3.0] nonenone [123] (Scheme 19).

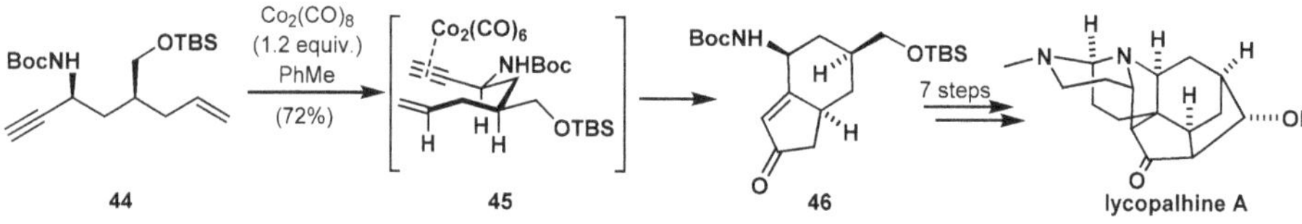

Scheme 19. Trauner's expedient synthesis of (+)-lycopalhine A.

Nakamura et al. have accomplished the stereoselective total synthesis of (+)-marrubiin involving a CyNH$_2$-promoted PKR and subsequent oxidative cleavage of the resultant cyclopentenone ring [124,125]. According to their DFT studies, the irreversible olefin insertion step is critical to the stereochemistry of PKR. The steric interaction could be avoided through a trans-fused chair–boat-like TS and therefore the exclusive isomer **48** was afforded (Scheme 20).

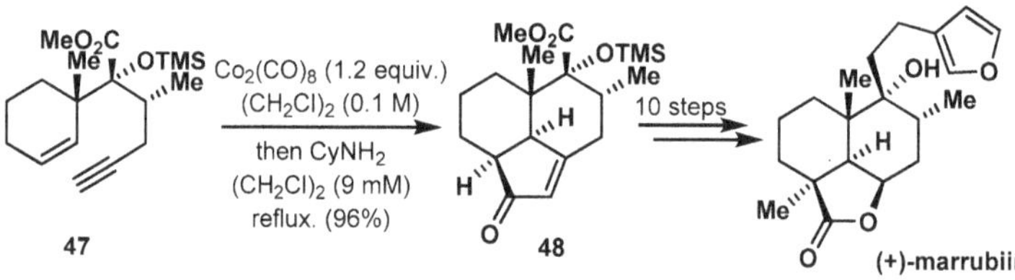

Scheme 20. Nakamura's total synthesis of (+)-marrubiin and (−)-marrulibacetal.

Yu et al. have developed an Rh(I)-catalyzed [3 + 2 + 1] cycloaddition of 1-ene/yne–vinylcyclopropanes (VCPs) and CO, which was used to construct 5,6-bicyclic advanced intermediate **50** from yne-VCP (±)-**49** [126]. The advanced intermediate **50** can be transformed into Gao's intermediate for the formal synthesis of gracilamine [127]. This cycloaddition provided a solution to construct the bridgehead quaternary carbon center. The diastereoselectivity was realized by the repulsion between the OTBS (TBS = *t*-butyldimethylsilyl) group and the vinyl moiety (Scheme 21).

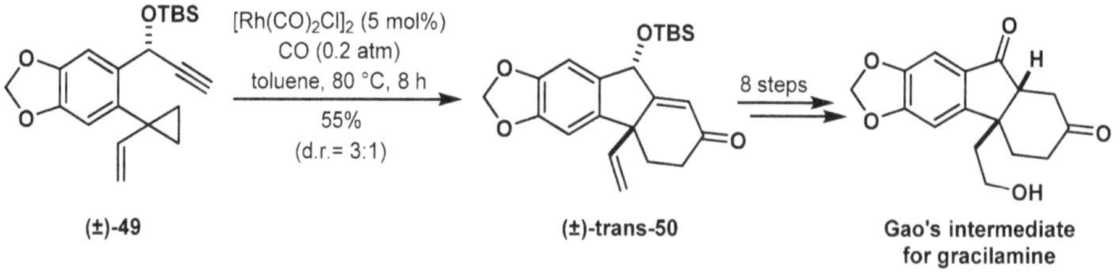

Scheme 21. Yu's formal synthesis of gracilamine.

In the synthesis of calcitriol, the active form of vitamin D$_3$, Mourino et al. utilized the NMO promoted PKR to form the 5,6-bicyclic core **52** in a diastereoselective way [128]. Intermediate **52** underwent Si-assisted allylic substitution and some other transformations to complete the synthesis of calcitriol (Scheme 22).

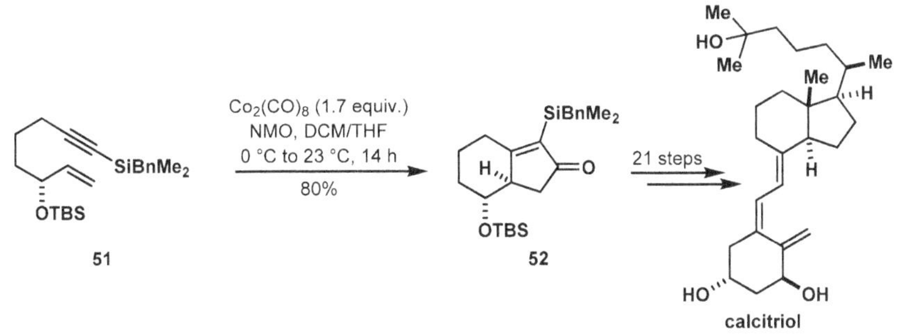

Scheme 22. Mourino's total synthesis of calcitriol.

Khan et al. delineated the collective total synthesis of iridolactones [129]. The newly constructed iridoid framework **54** was accomplished by a diastereoselective intramolecular PKR [130]. With the key intermediate **55** in hand, they demonstrated a general and simple route to access structurally divergent iridolactones (Scheme 23).

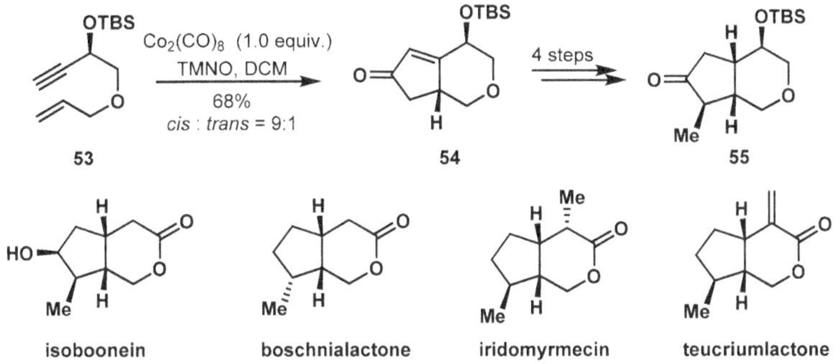

Scheme 23. Khan's total synthesis of several iridolactones.

A fawcettimine-type Lycopodium alkaloid (+)-sieboldine A contains an unprecedented fused tetracyclic skeleton and has been found to inhibit acetylcholinesterase with an IC_{50} value of 2.0 µM [131]. Mukai et al. have applied PKR to afford the bicyclo [4.3.0] nonenone derivative **57** with high stereoselectivity with an *ee* value of 93% in their total synthesis of (+)-sieboldine A [132] (Scheme 24).

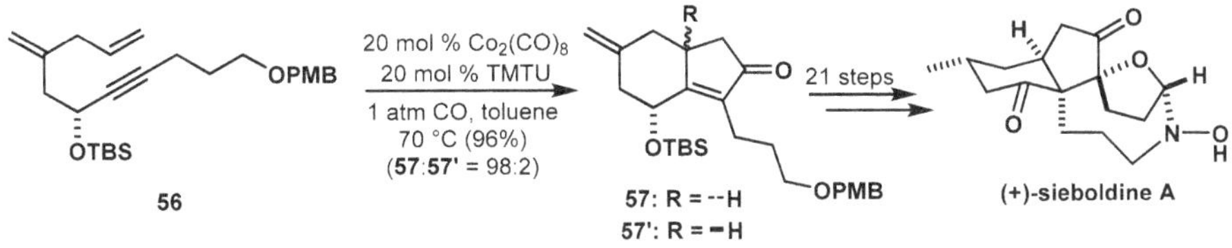

Scheme 24. Mukai's enantioselective total synthesis of (+)-sieboldine A.

Since the hPKR variant is much less reported, Zhai et al. applied an interesting hPKR in the formal synthesis of (±)-aplykurodinone-1 [133]. The tricyclic framework **59** has been constructed with a 60% yield through expeditiously one-pot intramolecular hPKR followed by the desilylation sequence. The hPKR is relatively rare to be found in natural product synthesis, and this application provided worthwhile insights for expanding the scope and boundaries (Scheme 25).

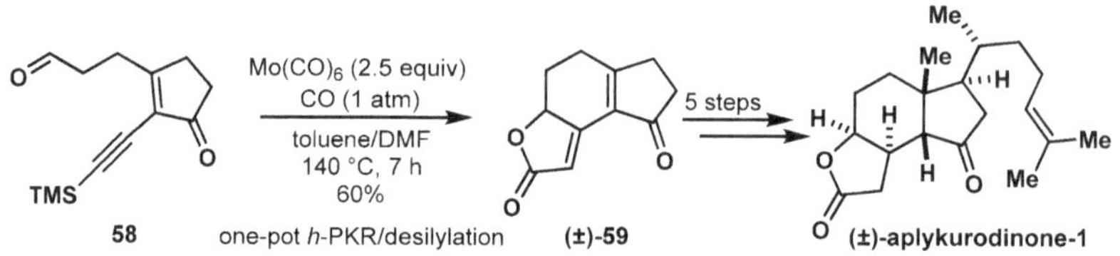

Scheme 25. Zhai's formal synthesis of (±)-aplykurodinone-1.

(−)-Sinoracutine, isolated from *Stephania cepharantha* in 2010 [134], proves to be a promising template for new neuroprotective reagents intervention as it was shown to increase cell viability against hydrogen peroxide-induced damage in PC12 cells [135]. Structurally, it features an unprecedented tetracyclic 6/6/5/5 skeleton that bears an N-methylpyrrolidine ring fused to acyclopentenone. In the first total synthesis of (−)-sinoracutine [136], Trauner et al. utilized intramolecular PKR under the oxidative condition as a key transformation to construct the tricycle product **61** from an enyne precursor **60**. The reaction was carried out in the presence of N-oxide dihydrate together with $Co_2(CO)_8$. The resulting tricyclic product **61** allows the concise total synthesis of (−)-sinoracutine with several steps of transformations, including a Mandai–Claisen reaction to install the quaternary stereocenter (Scheme 26).

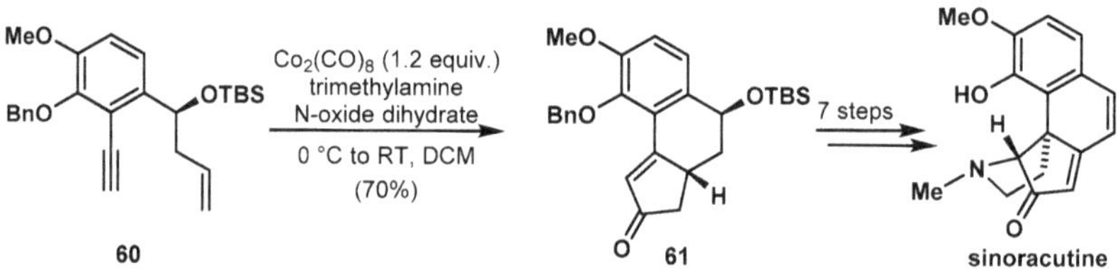

Scheme 26. Trauner's enantioselective synthesis and racemization of (−)-sinoracutine.

Cyanthiwigin type diterpenes are biologically important marine natural products mostly isolated from marine sponges *Epipolasis reiswigi* and *Mermekioderma styx*. Particularly, cyanthiwigin C and F show medium cytotoxicity against A549 cell lines [137]. In 2019, Yang et al. reported the total synthesis of 5-epi-cyanthiwigin I [138]. The key [5–6–7] tricarbocyclic fused core structure was constructed via a well-orchestrated Co-mediated intramolecular PKR, which has two cis-configured all-carbon quaternary chiral centers and an isopropyl group. Enyne **62** could be transformed into the tricyclic product **63** as the sole isomer in a 70% yield in the presence of a stoichiometric amount of $Co_2(CO)_8$ combined with NMO as the additive (Scheme 27).

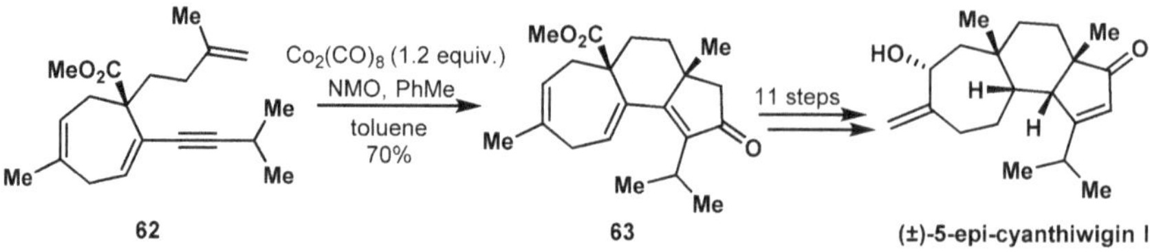

Scheme 27. Yang's stereoselective total synthesis of (±)-5-epi-cyanthiwigin I.

Lycopodium alkaloids are neuropharmacologically valuable scaffolds for central nervous system drug discovery. Takayama et al. reported an asymmetric total synthesis of lycopoclavamine A via a strategy involving a stereoselective PKR and a stereoselective conjugate addition to construct a quaternary carbon center at C12 [139]. The cobalt-mediated intramolecular PKR afforded a desired bicyclic enone **65** in a high yield as well as good diastereoselectivity (Scheme 28).

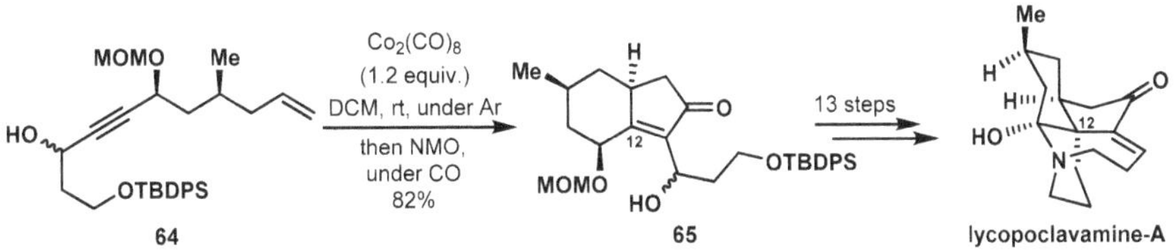

Scheme 28. Takayama's asymmetric total synthesis of lycopoclavamine-A.

Complex sesterterpenoids astellatol and astellatene were isolated from *Aspergillus stellatus* in 1989 [140], which feature highly congested and unusual pentacyclic skeletons and contain a unique bicyclo[4.1.1]octane moiety consisting of ten stereocenters and a cyclobutane containing two quaternary centers. In the total syntheses of (+)-astellatol and (−)-astellatene reported by Xu et al. [141], an intramolecular PKR was exploited to construct the 6,5-bicyclic core embedded in the right-wing scaffold. The desired hydrindane skeleton **67** was generated from enyne **66** with a promising yield and diastereoselectivity at the C7 quaternary carbon center (Scheme 29).

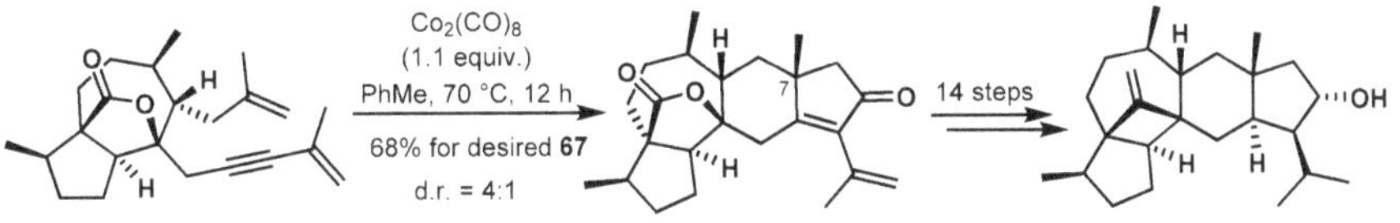

Scheme 29. Xu's asymmetric total synthesis of (+)-astellatol.

Porée et al. reported an elegant synthesis of allosecurinine, utilizing the W(CO)$_6$-promoted oxa-hetero-Pauson–Khand reaction (oxa-hPKR) in the late stage. Despite a low yield, the results constituted the first example of applying the W(CO)$_6$ complex in hPKR, constructing tetracyclic securinega alkaloid featuring an α, β-unsaturated γ-lactone moiety [142] (Scheme 30).

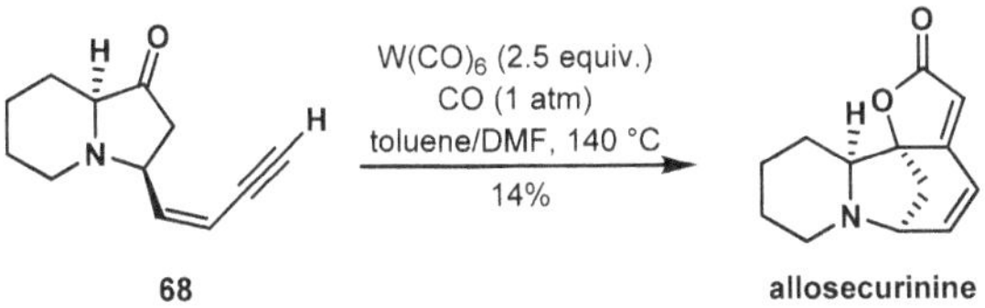

Scheme 30. Porée's enantioselective synthesis of (−)-allosecurinine.

The calyciphylline B-type alkaloids with a unique hexacyclic framework exhibited a variety of important pharmacological potentials. In the synthesis of daphlongamine H, Sarpong et al. used a late-stage cobalt-mediated PKR to accomplish the 6,5-bicyclic segment. The R configuration of C6 in the PKR enabled the desired 10-H α orientation in the PK product **70** [143] (Scheme 31).

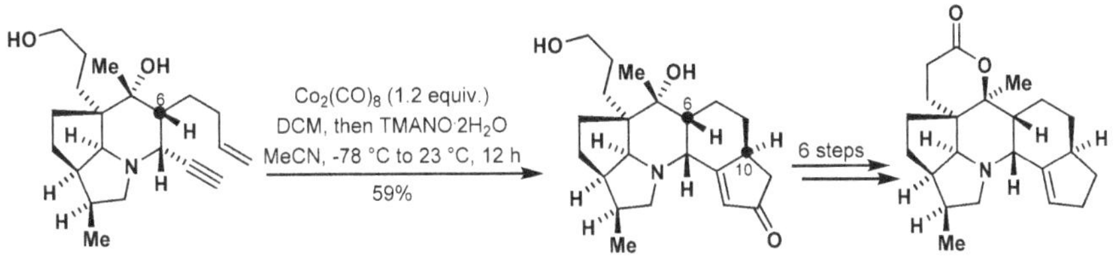

Scheme 31. Sarpong's total synthesis of (−)-daphlongamine H.

2.3. Construction of 5,7 Bicyclic Ring Systems

Thapsigargin (Tg1) and its analogs are biologically important candidates as potent inhibitors of the SERCA-pump protein, with the potential of application in a variety of medicinal areas [144,145]. Numerous attempts have been reported on the total synthesis of this bioactive molecule [146–148]. In 2019, Sorin et al. developed a linear route towards the core of Tg1, which features an allene-yne Rh(I)-catalyzed Pauson-Khand annulation (APKR) as key transformation [149]. The allene-yne precursor was generated from chiral propargylic alcohol **71**, which underwent a Ti(II) mediated reductive coupling to form diol **72**. The allene-yne product **73** was elaborated in several steps. The central feature was identified to be the Rh(I)-catalyzed Pauson-Khand annulation (APKR), resulting in the efficient synthesis of the Tg 1 framework bearing an enol ether moiety in a 71% yield (Scheme 32).

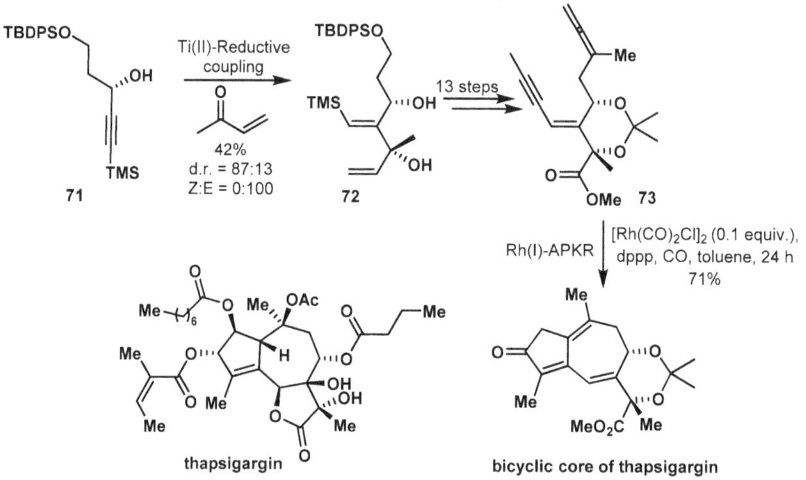

Scheme 32. Sorin's synthesis of a thapsigargin core.

Bufospirostenin A, isolated in 2017 from the toad *Bufo bufo gargarizans*, is an unusual steroid with rearranged A/B rings, possessing a cardioactive effect and promoting blood circulation through causing a 43% inhibition of Na/K ATPase (NKA) at 25 µM [150]. Very recently, a unique intramolecular rhodium-catalyzed PKR of an alkoxyallene-yne substrate was applied to construct the key [5–7] A-B ring system in the first total synthesis of bufospirostenin A reported by Li et al. [151]. Generated from Hajos-Parish ketone, alkyne **74** underwent 1,2-addition to afford precursor **75**, which further yielded tetracyclic product **76** catalyzed by [RhCl(CO$_2$)]$_2$ in the presence of a balloon pressure of CO in toluene with a high yield (85%). This work represented the first example of an intramolecular Pauson−Khand reaction of an alkoxyallene-yne in natural product synthesis (Scheme 33).

2.4. Construction of Macrocycles

The synthesis of macrocyclic natural products and related structures through a direct C-C bond formation is challenging. Widely applied methodologies include ring-closing metathesis (RCM), Nozaki-Hiyama-Kishi (NHK) reaction, and intramolecular Diels-Alder reactions. PKR has been used and confined in the synthesis of medium-sized rings (up to 11 atoms) [152,153], thus not yet been extended to macrocycles (Scheme 34).

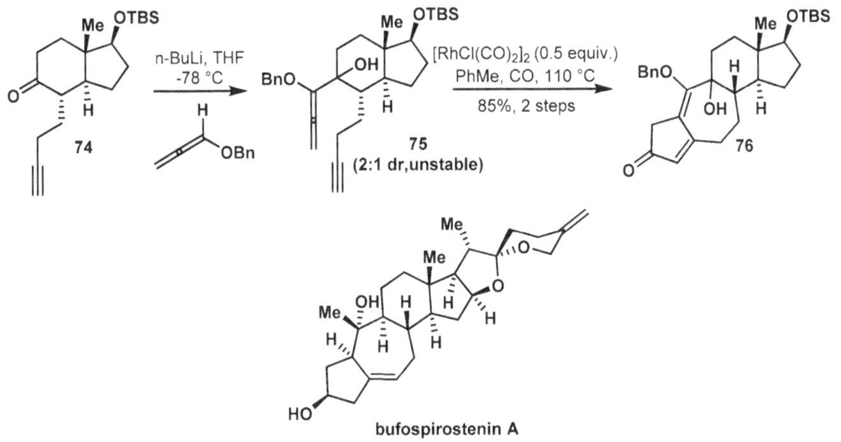

Scheme 33. Li's asymmetric total synthesis of bufospirostenin A.

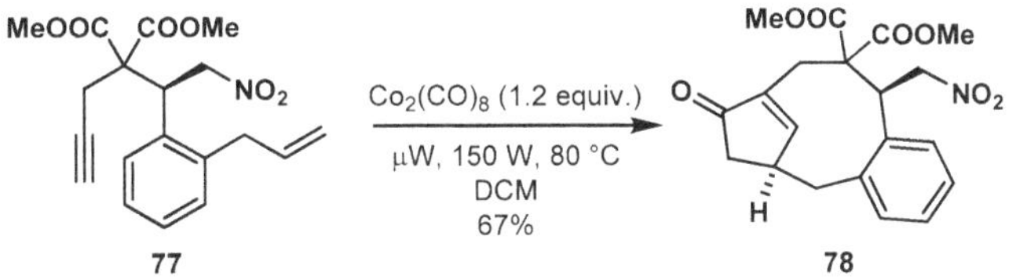

Scheme 34. Synthetic strategy for medium-sized rings.

Spring et al. reported their investigation of PKR for macrocyclization of a template substrate 79 [154]. After fluorous solid-phase extraction (F-SPE), optimized PKR conditions produced a mixture of structurally unusual macrocycles containing a cyclopentenone motif; these can be separated by HPLC, but they used the mixture in the modified phase (Scheme 35).

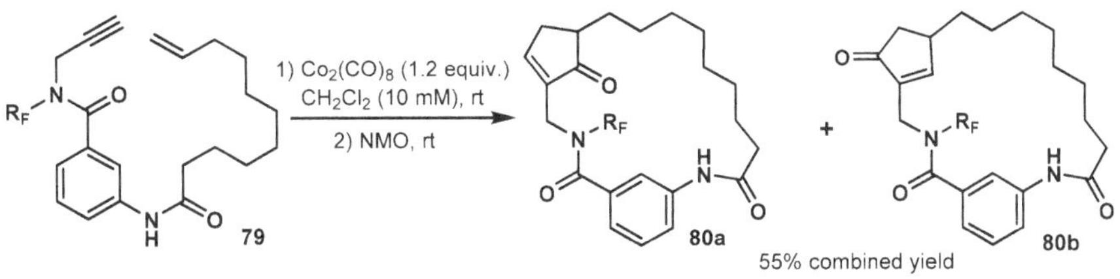

Scheme 35. Spring's synthetic strategy for structurally diverse and complex macrocycles.

3. Summary and Outlook

The extraordinary impact of the Pauson-Khand reaction on synthetic methods is still recognized nowadays, and attempts are currently undertaken to further extend the use of various metal-assisted chemistry to environmentally friendly processes within the strongly invoked green chemistry paradigm. The PKR, especially when conducted in an intramolecular fashion, has been widely used as a convenient and powerful tool for the construction of cyclopentenone structural units in natural product synthesis. Though the classical PK cycloaddition has the shortcoming that requires high temperatures and a long reaction time, chemists have developed a range of promoters (TMTU/NMO/TMAO = trimethylamine N-oxide, etc.) to circumvent this situation. Moreover, PKR has the merit of well tolerance to a broad variety of functional groups, such as alcohols, ethers, thioethers, esters, nitriles, amines, amides, sulfonamides, etc. With the impressive developments in the catalytic version of the Pauson–Khand reaction, the application will be more facilitated. Additionally, 4,5-fused bicycles afforded by intermolecular PKR patterns are still called for intensive studies.

Author Contributions: Conceptualization, S.C. and C.J.; data collection S.C. and C.J.; writing—original draft preparation, S.C. and C.J.; writing—review and editing, S.C., C.J., N.Z., Z.Y. and L.S. All authors have read and agreed to the published version of the manuscript.

Funding: This research was funded by the National Key Research and Development Program of China (grant no. 2018YFC0310905); National Science Foundation of China (grant nos. 21632002 and 21801123); and Shenzhen Basic Research Program (grant nos. 2019SHIBS0004, JCYJ20170818090044432, and JCYJ20180302180215524).

Acknowledgments: Research activities in related areas in Yang's lab are financially supported by the National Key Research and Development Program of China (grant no. 2018YFC0310905); National Science Foundation of China (grant nos. 21632002 and 21801123); and Shenzhen Basic Research Program (grant nos. 2019SHIBS0004, JCYJ20170818090044432, and JCYJ20180302180215524).

Conflicts of Interest: The authors declare no conflict of interest.

References

1. D'yakonov, V.A.; Trapeznikova, O.A.; de Meijere, A.; Dzhemilev, U.M. Metal Complex Catalysis in the Synthesis of Spirocarbocycles. *Chem. Rev.* **2014**, *114*, 5775–5814. [CrossRef] [PubMed]
2. Porcheddu, A.; Colacino, E.; De Luca, L.; Delogu, F. Metal-Mediated and Metal-Catalyzed Reactions Under Mechanochemical Conditions. *ACS Catal.* **2020**, *10*, 8344–8394. [CrossRef]
3. Rodríguez, J.; Martínez, C.M. Transition-Metal-Mediated Modification of Biomolecules. *Chem.-A Eur. J.* **2020**, *26*, 9792–9813. [CrossRef] [PubMed]
4. Khand, I.U.; Knox, G.R.; Pauson, P.L.; Watts, W.E.; Foreman, M.I. Organocobalt complexes. II. Reaction of acetylenehexacarbonyl dicobalt complexes, $(RC_2R_1) Co_2(CO)_6$, with norbornene and its derivatives. *J. Chem. Soc. Perkin Trans. 1* **1973**, *9*, 977–981. [CrossRef]
5. Chung, Y.K. Transition metal alkyne complexes: The Pauson–Khand reaction. *Coord. Chem. Rev.* **1999**, *188*, 297–314. [CrossRef]
6. Gibson, S.E.; Stevenazzi, A. The Pauson–Khand Reaction: The Catalytic Age Is Here. *Angew. Chem. Int. Ed.* **2003**, *42*, 1800–1810. [CrossRef] [PubMed]
7. Urgoiti, J.B.; Anorbe, L.; Serrano, L.P.; Dominguez, G.; Castells, L.P. The Pauson–Khand reaction, a powerful synthetic tool for the synthesis of complex molecules. *Chem. Soc. Rev.* **2004**, *33*, 32–42. [CrossRef]
8. Gibson, S.E.; Mainolfi, N. The Intermolecular Pauson–Khand Reaction. *Angew. Chem. Int. Ed.* **2005**, *44*, 3022–3037. [CrossRef]
9. Kitagaki, S.; Inagaki, F.; Mukai, C. [2+2+1] Cyclization of allenes. *Chem. Soc. Rev.* **2014**, *43*, 2956–2978. [CrossRef]
10. Lee, H.-W.; Kwong, F.-Y. A Decade of Advancements in Pauson–Khand-Type Reactions. *Eur. J. Org. Chem.* **2010**, *2010*, 789–811. [CrossRef]
11. Shi, L.; Yang, Z. Exploring the Complexity-Generating Features of the Pauson–Khand Reaction from a Synthetic Perspective. *Eur. J. Org. Chem.* **2016**, *2016*, 2356–2368. [CrossRef]
12. Ricker, J.D.; Geary, L.M. Recent Advances in the Pauson–Khand Reaction. *Top. Catal.* **2017**, *60*, 609–619. [CrossRef] [PubMed]
13. Jeong, N.; Hwang, S.H.; Lee, Y.S.; Chung, Y.K. Catalytic Version of the Intramolecular Pauson-Khand Reaction. *J. Am. Chem. Soc.* **1994**, *116*, 3159–3160. [CrossRef]
14. Hayashi, M.; Hashimoto, Y.; Yamamoto, Y.; Usuki, J.; Saigo, K. Phosphane Sulfide/Octacarbonyldicobalt-Catalyzed Pauson-Khand Reaction under an Atmospheric Pressure of Carbon Monoxide. *Angew. Chem. Int. Ed.* **2000**, *39*, 631–632. [CrossRef]
15. Krafft, M.E.; Boñaga, L.V.R.; Wright, J.A.; Hirosawa, C. Cobalt Carbonyl-Mediated Carbocyclizations of Enynes: Generation of Bicyclooctanones or Monocyclic Alkenes. *J. Org. Chem.* **2002**, *67*, 1233–1246. [CrossRef]
16. Wang, Y.; Xu, L.; Yu, R.; Chen, J.; Yang, Z. $CoBr_2$–TMTU–zinc catalysed-Pauson–Khand reaction. *Chem. Commun.* **2012**, *48*, 8183–8185. [CrossRef]
17. Orgue, S.; Leon, T.; Riera, A.; Verdaguer, X. Asymmetric Intermolecular Cobalt Catalyzed Pauson-Khand Reaction Using a Pstereogenic BisPhosphane. *Org. Lett.* **2015**, *17*, 250–253. [CrossRef]

18. Koga, Y.; Kobayashi, T.; Narasaka, K. Rhodium-Catalyzed Intramolecular Pauson-Khand Reaction. *Chem. Lett.* **1998**, *3*, 249–250. [CrossRef]
19. Jeong, N.; Lee, S.; Sung, B.K. Rhodium(I)-Catalyzed Intramolecular Pauson−Khand Reaction. *Organometallics* **1998**, *17*, 3642–3644. [CrossRef]
20. Fan, B.M.; Xie, J.H.; Li, S.; Tu, Y.Q.; Zhou, Q.L. Rhodium-Catalyzed Asymmetric Pauson–Khand Reaction Using Monophosphoramidite Ligand SIPHOS. *Adv. Synth. Catal.* **2005**, *347*, 759–762. [CrossRef]
21. Jeong, N.; Sung, B.K.; Choi, Y.K. Rhodium(I)-Catalyzed Asymmetric Intramolecular Pauson-Khand-Type Reaction. *J. Am. Chem. Soc.* **2000**, *122*, 6771–6772. [CrossRef]
22. Jeong, N.; Kim, D.H.; Choi, J.H. Desymmetrization of meso-dienyne by asymmetric Pauson–Khand type reaction catalysts. *Chem. Commun.* **2004**, 1134–1135. [CrossRef] [PubMed]
23. Kwong, F.Y.; Li, Y.M.; Lam, W.H.; Qiu, L.Q.; Lee, H.W.; Yeung, C.H.; Chan, K.S.; Chan, A.S.C. Rhodium-Catalyzed Asymmetric Aqueous Pauson–Khand-Type Reaction. *Chem. Eur. J.* **2005**, *11*, 3872–3880. [CrossRef]
24. Chen, G.Q.; Shi, M. Rhodium-catalyzed tandem Pauson–Khand type reactions of 1,4-enynes tethered by a cyclopropyl group. *Chem. Commun.* **2013**, *49*, 698–700. [CrossRef] [PubMed]
25. Wang, H.J.; Sawyer, J.R.; Evans, P.A.; Baik, M.H. Mechanistic Insight into the Diastereoselective Rhodium-Catalyzed Pauson–Khand Reaction: Role of Coordination Number in Stereocontrol. *Angew. Chem. Int. Ed.* **2008**, *47*, 342–345. [CrossRef]
26. Kondo, T.; Suzuki, N.; Okada, T.; Mitsudo, T. First Ruthenium-Catalyzed Intramolecular Pauson-Khand Reaction. *J. Am. Chem. Soc.* **1997**, *119*, 6187–6188. [CrossRef]
27. Morimoto, T.; Chatani, N.; Fukumoto, Y.; Murai, S. $Ru_3(CO)_{12}$-Catalyzed Cyclocarbonylation of 1,6-Enynes to Bicyclo[3.3.0]Octenones. *J. Org. Chem.* **1997**, *62*, 3762–3765. [CrossRef]
28. Chatani, N.; Morimoto, T.; Fukumoto, Y.; Murai, S. $Ru_3(CO)_{12}$-Catalyzed Cyclocarbonylation of Yne-Aldehydes to Bicyclic α, β-Unsaturated γ-Butyrolactones. *J. Am. Chem. Soc.* **1998**, *120*, 5335–5336. [CrossRef]
29. Miura, H.; Takeuchi, K.; Shishido, T. Intermolecular [2+2+1] Carbonylative Cycloaddition of Aldehydes with Alkynes, and Subsequent Oxidation to γ-Hydroxybutenolides by a Supported Ruthenium Catalyst. *Angew. Chem. Int. Ed.* **2016**, *55*, 278–282. [CrossRef]
30. Kondo, T.; Nomura, M.; Ura, Y.; Wada, K.; Mitsudo, T. Ruthenium-catalyzed [2 + 2 + 1] Cocyclization of Isocyanates, Alkynes, and CO Enables the Rapid Synthesis of Polysubstituted Maleimides. *J. Am. Chem. Soc.* **2006**, *128*, 14816–14817. [CrossRef]
31. Hicks, F.A.; Kablaoui, N.M.; Buchwald, S.L. Scope of the Intramolecular Titanocene-Catalyzed Pauson-Khand Type Reaction. *J. Am. Chem. Soc.* **1999**, *121*, 5881–5898. [CrossRef]
32. Hicks, F.A.; Buchwald, S.L. An intramolecular Titanium-Catalyzed Asymmetric Pauson-Khand Type Reaction. *J. Am. Chem. Soc.* **1999**, *121*, 7026–7033. [CrossRef]
33. Zhao, Z.B.; Ding, Y.; Zhao, G. Bicyclization of Enynes Using the Cp_2TiCl_2–Mg–BTC System: A Practical Method to Bicyclic Cyclopentenones. *J. Org. Chem.* **1998**, *63*, 9285–9291. [CrossRef]
34. Hicks, F.A.; Kablaoui, N.M.; Buchwald, S.L. Titanocene-Catalyzed Cyclocarbonylation of Enynes to Cyclopentenones. *J. Am. Chem. Soc.* **1996**, *118*, 9450–9451. [CrossRef]
35. Shibata, T.; Takagi, K. Iridium-Chiral Diphosphine Complex Catalyzed Highly Enantioselective Pauson-Khand-Type Reaction. *J. Am. Chem. Soc.* **2000**, *122*, 9852–9853. [CrossRef]
36. Shibata, T.; Toshida, N.; Yamasaki, M.; Maekawa, S.; Takagi, K. Iridium-catalyzed enantioselective Pauson–Khand-type reaction of 1,6-enynes. *Tetrahedron* **2005**, *61*, 9974–9979. [CrossRef]
37. Kwong, F.Y.; Lee, H.W.; Lam, W.H.; Qiu, L.Q.; Chan, A.S.C. Iridium-catalyzed cascade decarbonylation/highly enantioselective Pauson–Khand-type cyclization reactions. *Tetrahedron Asymm.* **2006**, *17*, 1238–1252. [CrossRef]
38. Zhang, M.H.; Buchwald, S.L. A Nickel(0)-Catalyzed Process for the Transformation of Enynes to Bicyclic Cyclopentenones. *J. Org. Chem.* **1996**, *61*, 4498–4499. [CrossRef]
39. Kent, J.L.; Wan, H.H.; Brummond, K.M. A new allenic Pauson-Khand cycloaddition for the preparation of α-methylene cyclopentenones. *Tetrahedron Lett.* **1995**, *36*, 2407–2410. [CrossRef]
40. Adrio, J.; Rivero, M.R.; Carretero, J.C. Mild and Efficient Molybdenum-Mediated Pauson–Khand-Type Reaction. *Org. Lett.* **2005**, *7*, 431–434. [CrossRef]

41. Shibata, T.; Koga, Y.; Narasaka, K. Intra- and Intermolecular Allene–Alkyne Coupling Reactions by the Use of Fe(CO)$_4$(NMe$_3$). *Bull. Chem. Soc. Jpn.* **1995**, *68*, 911–919. [CrossRef]
42. Rutherford, D.T.; Christie, S.D.R. Soluble polymer-supported synthesis of arylpiperazines. *Tetrahedron Lett.* **1998**, *39*, 9505–9508.
43. Park, K.H.; Son, S.U.; Chung, Y.K. Soluble polymer-supported synthesis of arylpiperazines. *Chem. Commun.* **2003**, 1898–1899. [CrossRef] [PubMed]
44. Park, K.H.; Son, S.U.; Chung, Y.K. Immobilized heterobimetallic Ru/Co nanoparticle-catalyzed Pauson–Khand-type reactions in the presence of pyridylmethyl formate. *Chem. Commun.* **2008**, 2388–2390. [CrossRef] [PubMed]
45. Khand, I.U.; Pauson, P.L. Organometallic Route to 2,7-Dihydrothiepin-1,1-Dioxides. *Heterocycles* **1978**, *11*, 59–67.
46. Magnus, P.; Principe, L.M. Origins of 1,2-Stereoselectivity and 1,3-Stereoselectivity in Dicobaltoctacarbonyl Alkene Alkyne Cyclizations for the Synthesis of Substituted Bicyclo[3.3.0]Octenones. *Tetrahedron Lett.* **1985**, *26*, 4851–4854. [CrossRef]
47. Magnus, P.; Exon, C.; Albaughrobertson, P. Dicobaltoctacarbonyl Alkyne Complexes as Intermediates in the Synthesis of Bicyclo[3.3.0]Octenones for the Synthesis of Coriolin and Hirsutic Acid. *Tetrahedron* **1985**, *41*, 5861–5869. [CrossRef]
48. Magnus, P.; Principe, L.M.; Slater, M.J. Stereospecific Dicobalt Octacarbonyl Mediated Enyne Cyclization for the Synthesis of the Cytotoxic Sesquiterpene (+/−)-Quadrone. *J. Org. Chem.* **1987**, *52*, 1483–1486. [CrossRef]
49. Labelle, B.E.; Knudsen, M.J.; Olmstead, M.M.; Hope, H.; Yanuck, M.D.; Schore, N.E. Synthesis of 11-Oxatricyclo[5.3.1.02,6] Undecane Derivatives Via Organometallic Cyclizations. *J. Org. Chem.* **1985**, *50*, 5215–5222. [CrossRef]
50. Krafft, M.E.; Romero, R.H.; Scott, I.L. Pauson-Khand Reaction with Electron-Deficient Alkynes. *J. Org. Chem.* **1992**, *57*, 5277–5278. [CrossRef]
51. Robert, F.; Milet, A.; Gimbert, Y.; Konya, D.; Greene, A.E. Regiochemistry in the Pauson-Khand Reaction: Has a Trans Effect Been Overlooked? *J. Am. Chem. Soc.* **2001**, *123*, 5396–5400. [CrossRef]
52. Yamamoto, Y.; Kuwabara, S.; Ando, Y.; Nagata, H.; Nishiyama, H.; Itoh, K. Palladium(0)-Catalyzed Cyclization of Electron-Deficient Enynes and Enediynes. *J. Org. Chem.* **2004**, *69*, 6697–6705. [CrossRef] [PubMed]
53. Pauson, P.L. The Khand Reaction—A Convenient and General-Route to a Wide-Range of Cyclopentenone Derivatives. *Tetrahedron* **1985**, *41*, 5855–5860. [CrossRef]
54. De Bruin, T.J.M.; Milet, A.; Greene, A.E.; Gimbert, Y. Insight into the Reactivity of Olefins in the Pauson-Khand Reaction. *J. Org. Chem.* **2004**, *69*, 1075–1080. [CrossRef] [PubMed]
55. Krafft, M.E. Regiocontrol in the Intermolecular Cobalt-Catalyzed Olefin Acetylene Cyclo-Addition. *J. Am. Chem. Soc.* **1988**, *110*, 968–970. [CrossRef]
56. Krafft, M.E. Steric Control in the Pauson Cyclo-Addition—Further Support for the Proposed Mechanism. *Tetrahedron Lett.* **1988**, *29*, 999–1002. [CrossRef]
57. Sola, J.; Riera, A.; Verdaguer, X.; Maestro, M.A. Phosphine-Substrate Recognition through the C-H···O Hydrogen Bond: Application to the Asymmetric Pauson-Khand Reaction. *J. Am. Chem. Soc.* **2005**, *127*, 13629–13633. [CrossRef]
58. Ahmar, M.; Antras, F.; Cazes, B. Pauson-Khand Reaction with Allenic Compounds.1. Synthesis of 4-Alkylidene-2-Cyclopentenones. *Tetrahedron Lett.* **1995**, *36*, 4417–4420. [CrossRef]
59. Ahmar, M.; Chabanis, O.; Gauthier, J.; Cazes, B. Pauson-Khand Reaction with Allenic Compounds II: Reactivity of Functionalized Allenes. *Tetrahedron Lett.* **1997**, *38*, 5277–5280. [CrossRef]
60. Wu, N.; Deng, L.J.; Liu, L.Z.; Liu, Q.; Li, C.C.; Yang, Z. Reverse Regioselectivity in the Palladium(II) Thiourea Catalyzed Intermolecular Pauson-Khand Reaction. *Chem. Asian J.* **2013**, *8*, 65–68. [CrossRef]
61. Smit, W.A.; Gybin, A.S.; Shashkov, A.S.; Strychkov, Y.T.; Kyzmina, L.G.; Mikaelian, G.S.; Caple, R.; Swanson, E.D. New Route to the Synthesis of Polycyclic Compounds Based on a Stepwise Ade-Reaction of Dicobalt Hexacarbonyl Complexes of Conjugated Enynes with a Subsequent Intramolecular Khand-Pauson Type Reaction. *Tetrahedron Lett.* **1986**, *27*, 1241–1244. [CrossRef]
62. Shen, J.K.; Gao, Y.C.; Shi, Q.Z.; Basolo, F. Oxygen Atom Transfer-Reactions to Metal-Carbonyls—Kinetics and Mechanism of Co Substitution of Fe(Co)$_5$, Ru(Co)$_5$, Os(Co)$_5$ in the Presence of (CH$_3$)$_3$NO. *Organometallics* **1989**, *8*, 2144–2147. [CrossRef]

63. Shambayati, S.; Crowe, W.E.; Schreiber, S.L. N-Oxide Promoted Pauson-Khand Cyclizations at Room-Temperature. *Tetrahedron Lett.* **1990**, *31*, 5289–5292. [CrossRef]
64. Gallagher, A.G.; Tian, H.; Torres-Herrera, O.A.; Yin, S.; Xie, A.; Lange, D.M.; Wilson, J.K.; Mueller, L.G.; Gau, M.R.; Carroll, P.J.; et al. Access to Highly Functionalized Cyclopentenones via Diastereoselective Pauson-Khand Reaction of Siloxy-Tethered 1,7-Enynes. *Org. Lett.* **2019**, *21*, 8646–8651. [CrossRef]
65. Hiroi, K.; Watanabe, T.; Kawagishi, R.; Abe, I. Asymmetric Catalytic Pauson-Khand Reactions with Chiral Phosphine Ligands: Dramatic Effects of Substituents in 1,6-Enyne Systems. *Tetrahedron Lett.* **2000**, *41*, 891–895. [CrossRef]
66. Tang, Y.F.; Deng, L.J.; Zhang, Y.D.; Dong, G.B.; Chen, J.H.; Yang, Z. Tetramethyl thiourea/$Co_2(CO)_8$-catalyzed Pauson-Khand Reaction under Balloon Pressure of CO. *Org. Lett.* **2005**, *7*, 593–595. [CrossRef] [PubMed]
67. Sugihara, T.; Yamada, M.; Ban, H.; Yamaguchi, M.; Kaneko, C. Rate Enhancement of the Pauson-Khand Reaction by Primary Amines. *Angew. Chem. Int. Ed.* **1997**, *36*, 2801–2804. [CrossRef]
68. Del Valle, C.P.; Milet, A.; Gimbert, Y.; Greene, A.E. Lewis Base Promoters in the Pauson-Khand Reaction: A Different Scenario. *Angew. Chem. Int. Ed.* **2005**, *44*, 5717–5719. [CrossRef]
69. Kobayashi, T.; Koga, Y.; Narasaka, K. The Rhodium-Catalyzed Pauson-Khand Reaction. *J. Organomet. Chem.* **2001**, *624*, 73–87. [CrossRef]
70. Schmid, T.M.; Consiglio, G. Mechanistic and Stereochemical Aspects of the Asymmetric Cyclocarbonylation of 1,6-Enynes with Rhodium Catalysts. *Chem. Commun.* **2004**, *20*, 2318–2319. [CrossRef]
71. Wender, P.A.; Deschamps, N.M.; Williams, T.J. Intermolecular Dienyl Pauson–Khand Reaction. *Angew. Chem. Int. Ed.* **2004**, *43*, 3076–3079. [CrossRef] [PubMed]
72. Wender, P.A.; Croatt, M.P.; Deschamps, N.M. Metal-Catalyzed [2+2+1] Cycloadditions of 1,3-Dienes, Allenes, and CO. *Angew. Chem. Int. Ed.* **2006**, *45*, 2459–2462. [CrossRef] [PubMed]
73. Shibata, T.; Toshida, N.; Takagi, K. Catalytic Pauson-Khand-Type Reaction Using Aldehydes as a CO Source. *Org. Lett.* **2002**, *4*, 1619–1621. [CrossRef] [PubMed]
74. Park, J.H.; Cho, Y.; Chung, Y.K. Rhodium-Catalyzed Pauson-Khand-Type Reaction Using Alcohol as a Source of Carbon Monoxide. *Angew. Chem. Int. Ed.* **2010**, *49*, 5138–5141. [CrossRef] [PubMed]
75. Lee, H.W.; Chan, A.S.C.; Kwong, F.Y. Formate as a CO surrogate for cascade processes: Rh-catalyzed cooperative decarbonylation and asymmetric Pauson–Khand-type cyclization reactions. *Chem. Commun.* **2007**, *25*, 2633–2635. [CrossRef] [PubMed]
76. Lang, X.-D.; You, F.; He, X.; Yu, Y.-C.; He, L.-N. Rhodium(I)-catalyzed Pauson Khand-type Reaction Using Formic Acid as a CO Surrogate: An Alternative Approach for Indirect CO_2 Utilization. *Green Chem.* **2019**, *21*, 509–514. [CrossRef]
77. Jorgensen, L.; McKerrall, S.J.; Kuttruff, C.A.; Ungeheuer, F.; Felding, J.; Baran, P.S. 14-Step Synthesis of (+)-Ingenol from (+)-3-Carene. *Science* **2013**, *341*, 878–882. [CrossRef] [PubMed]
78. Dai, M.J.; Liang, B.; Wang, C.H.; Chen, J.H.; Yang, Z. Synthesis of a Novel C-2-Symmetric Thiourea and Its Application in the Pd-Catalyzed Cross-Coupling Reactions with Arenediazonium Salts under Aerobic Conditions. *Org. Lett.* **2004**, *6*, 221–224. [CrossRef]
79. Xiong, Z.C.; Wang, N.D.; Dai, M.J.; Li, A.; Chen, J.H.; Yang, Z. Synthesis of Novel Palladacycles and Their Application in Heck and Suzuki Reactions under Aerobic Conditions. *Org. Lett.* **2004**, *6*, 3337–3340. [CrossRef]
80. Liao, Y.; Smith, J.; Fathi, R.; Yang, Z. Novel Pd(II)-Mediated Cascade Carboxylative Annulation to Construct BenzoFuran-3-Carboxylic Acids. *Org. Lett.* **2005**, *7*, 2707–2709. [CrossRef]
81. Deng, L.J.; Liu, J.; Huang, J.Q.; Hu, Y.H.; Chen, M.; Lan, Y.; Chen, J.H.; Lei, A.W.; Yang, Z. Effect of Lithium Chloride on Tuning the Reactivity of Pauson-Khand Reactions Catalyzed by Palladium-Tetramethylthiourea. *Synthesis* **2007**, *2007*, 2565–2570. [CrossRef]
82. Tang, Y.; Deng, L.; Zhang, Y.; Dong, G.; Chen, J.; Yang, Z. Thioureas as Ligands in the Pd-Catalyzed Intramolecular Pauson–Khand Reaction. *Org. Lett.* **2005**, *7*, 1657–1659. [CrossRef]
83. Lan, Y.; Deng, L.J.; Liu, J.; Wang, C.; Wiest, O.; Yang, Z.; Wu, Y.D. On the Mechanism of the Palladium Catalyzed Intramolecular Pauson-Khand-Type Reaction. *J. Org. Chem.* **2009**, *74*, 5049–5058. [CrossRef] [PubMed]
84. Crowe, W.E.; Vu, A.T. Direct Synthesis of Fused, Bicyclic γ-Butyrolactones via Tandem Reductive Cyclization–Carbonylation of Tethered Enals and Enones. *J. Am. Chem. Soc.* **1996**, *118*, 1557–1558. [CrossRef]

85. Mandal, S.K.; Amin, S.R.; Crowe, W.E. γ-Butyrolactone Synthesis via Catalytic Asymmetric Cyclocarbonylation. *J. Am. Chem. Soc.* **2001**, *123*, 6457–6458. [CrossRef] [PubMed]
86. Kablaoui, N.M.; Hicks, F.A.; Buchwald, S.L. Diastereoselective Synthesis of γ-Butyrolactones from Enones Mediated or Catalyzed by a Titanocene Complex. *J. Am. Chem. Soc.* **1996**, *118*, 5818–5819. [CrossRef]
87. Kablaoui, N.M.; Hicks, F.A.; Buchwald, S.L. Titanocene-Catalyzed Cyclocarbonylation of o-Allyl Aryl Ketones to γ-Butyrolactones. *J. Am. Chem. Soc.* **1997**, *119*, 4424–4431. [CrossRef]
88. Chatani, N.; Morimoto, T.; Kamitani, A.; Fukumoto, Y.; Murai, S.J. $Ru_3(CO)_{12}$-catalyzed reaction of yne–imines with carbon monoxide leading to bicyclic α, β-unsaturated lactams. *Organomet. Chem.* **1999**, *579*, 177–181. [CrossRef]
89. Adrio, J.; Carretero, J.C. Butenolide Synthesis by Molybdenum-Mediated Hetero-Pauson–Khand Reaction of Alkynyl Aldehydes. *J. Am. Chem. Soc.* **2007**, *129*, 778–779. [CrossRef]
90. Saito, T.; Sugizaki, K.; Osada, H.; Kutsumura, N.; Otani, T. A Hetero Pauson-Khand Reaction of Ketenimines: A New Synthetic Method for γ-Exomethylene-α, β-unsaturated γ-Lactams. *Heterocycles* **2010**, *80*, 207–211. [CrossRef]
91. Finnegan, D.F.; Snapper, M.L. Formation of Polycyclic Lactones through a Ruthenium-Catalyzed Ring-Closing Metathesis/Hetero-Pauson–Khand Reaction Sequence. *J. Org. Chem.* **2011**, *76*, 3644–3653. [CrossRef]
92. Gao, P.; Xu, P.F.; Zhai, H. Expeditious Construction of (+)-Mintlactone via Intramolecular Hetero-Pauson-Khand Reaction. *J. Org. Chem.* **2009**, *74*, 2592–2593. [CrossRef] [PubMed]
93. Chen, J.; Gao, P.; Yu, F.; Yang, Y.; Zhu, S.; Zhai, H. Total Synthesis of (±)-Merrilactone A. *Angew. Chem. Int. Ed.* **2012**, *51*, 5897–5899. [CrossRef] [PubMed]
94. Lu, H.-H.; Martinez, M.D.; Shenvi, R.A. An eight-step gram-scale synthesis of (−)-jiadifenolide. *Nat. Chem.* **2015**, *7*, 604–607. [CrossRef] [PubMed]
95. Chirkin, E.; Michel, S.; Porée, F.H. Viability of a [2 + 2 + 1] Hetero-Pauson–Khand Cycloaddition Strategy toward Securinega Alkaloids: Synthesis of the BCD-Ring Core of Securinine and Related Alkaloids. *J. Org. Chem.* **2015**, *80*, 6525–6528. [CrossRef] [PubMed]
96. Mukai, C.; Yoshida, T.; Sorimachi, M.; Odani, A. $Co_2(CO)_8$-Catalyzed Intramolecular Hetero-Pauson–Khand Reaction of Alkynecarbodiimide: Synthesis of (±)-Physostigmine. *Org. Lett.* **2006**, *8*, 83–86. [CrossRef] [PubMed]
97. Huang, Z.; Huang, J.; Qu, Y.; Zhang, W.; Gong, J.; Yang, Z. Total Syntheses of Crinipellins Enabled by Cobalt-Mediated and Palladium-Catalyzed Intramolecular Pauson-Khand Reactions. *Angew. Chem. Int. Ed.* **2018**, *57*, 8744–8748. [CrossRef]
98. Peng, C.; Arya, P.; Zhou, Z.; Snyder, S.A. A Concise Total Synthesis of (+)-Waihoensene Guided by Quaternary Center Analysis. *Angew. Chem. Int. Ed.* **2020**, *59*, 13521–13525. [CrossRef]
99. Qu, Y.; Wang, Z.; Zhang, Z.; Zhang, W.; Huang, J.; Yang, Z. Asymmetric Total Synthesis of (+)-Waihoensene. *J. Am. Chem. Soc.* **2020**, *142*, 6511–6515. [CrossRef]
100. Knudsen, M.J.; Schore, N.E. Synthesis of the Angularly Fused Triquinane Skeleton Via Intramolecular Organometallic Cyclization. *J. Org. Chem.* **1984**, *49*, 5025–5026. [CrossRef]
101. Pallerla, M.K.; Fox, J.M. Enantioselective Synthesis of (−)-Pentalenene. *Org. Lett.* **2007**, *9*, 5625–5628. [CrossRef] [PubMed]
102. Millham, A.B.; Kier, M.J.; Leon, R.M.; Karmakar, R.; Stempel, Z.D.; Micalizio, G.C. A Complementary Process to PausonKhand-Type Annulation Reactions for the Construction of Fully Substituted Cyclopentenones. *Org. Lett.* **2019**, *21*, 567–570. [CrossRef] [PubMed]
103. Bird, R.; Knipe, A.C.; Stirling, C.J.M.J. Intramolecular Reactions. Part X. Transition States in the Cyclisation of N-ω-Halogeno-alkylamines and –sulphonamides. *Chem. Soc. Perkin Trans. 2* **1973**, *9*, 1215–1220. [CrossRef]
104. Wiesner, K.; Valenta, Z.; Findlay, J.A. The structure of ryanodine. *Tetrahedron Lett.* **1967**, *8*, 221–223. [CrossRef]
105. Srivastava, S.N.; Przybylska, M. The molecular structure of ryanodol-p-bromo benzyl ether. *Can. J. Chem.* **1968**, *46*, 795–797. [CrossRef]
106. Chuang, K.V.; Xu, C.; Reisman, S.E. A 15-step synthesis of (+)-ryanodol. *Science* **2016**, *353*, 912–915. [CrossRef] [PubMed]
107. Zhang, Z.; Li, Y.; Zhao, D.; He, Y.; Gong, J.; Yang, Z. A Concise Synthesis of Presilphiperfolane Corethrough a Tandem TMTU-Co-Catalyzed Pauson-Khand Reaction and a 6pi Electrocyclization Reaction (TMTU = Tetramethyl Thiourea). *Chem. Eur. J.* **2017**, *23*, 1258–1262. [CrossRef] [PubMed]

108. Zhang, Z.; Zhao, D.; He, Y.; Yang, Z.; Gong, J. Total syntheses of dehydrobotrydienal, dehydrobotrydienol and 10-oxodehydro- dihydrobotrydial. *Chin. Chem. Lett.* **2019**, *30*, 1503–1505. [CrossRef]
109. Zhang, Z.; Zhao, D.; Zhang, Z.; Tan, X.; Gong, J.; Yang, Z. Synthesis of 4-Desmethyl-rippertenol and 7-Epi-rippertenol via Photo-induced Cyclization of Dienones. *CCS Chem.* **2020**. [CrossRef]
110. Sun, T.W.; Liu, D.D.; Wang, K.Y.; Tong, B.Q.; Xie, J.X.; Jiang, Y.L.; Li, Y.; Zhang, B.; Liu, Y.F.; Wang, Y.X.; et al. Asymmetric Total Synthesis of Lancifodilactone G Acetate. 1. Diastereoselective Synthesis of CDEFGH Ring System. *J. Org. Chem.* **2018**, *83*, 6893–6906. [CrossRef]
111. Wang, K.Y.; Liu, D.D.; Sun, T.W.; Lu, Y.; Zhang, S.L.; Li, Y.H.; Han, Y.X.; Liu, H.Y.; Peng, C.; Wang, Q.Y.; et al. Asymmetric Total Synthesis of Lancifodilactone G Acetate. 2. Final Phase and Completion of the Total Synthesis. *J. Org. Chem.* **2018**, *83*, 6907–6923. [CrossRef] [PubMed]
112. Zhang, W.; Ding, M.; Li, J.; Guo, Z.; Lu, M.; Chen, Y.; Liu, L.; Shen, Y.H.; Li, A. Total Synthesis of Hybridaphniphylline B. *J. Am. Chem. Soc.* **2018**, *140*, 4227–4231. [CrossRef] [PubMed]
113. Chung, Y.K.; Lee, B.Y.; Jeong, N.; Hudecek, M.; Pauson, P.L. Promoters for the (alkyne) hexacarbonyldicobalt-based cyclopentenone synthesis. *Organometallics* **1993**, *12*, 220–223. [CrossRef]
114. Hu, N.; Dong, C.; Zhang, C.; Liang, G. Total Synthesis of (−)-Indoxamycins A and B. *Angew. Chem. Int. Ed.* **2019**, *58*, 6659–6662. [CrossRef]
115. Hu, X.; Musacchio, A.J.; Shen, X.; Tao, Y.; Maimone, T.J. Allylative Approaches to the Synthesis of Complex Guaianolide Sesquiterpenes from Apiaceae and Asteraceae. *J. Am. Chem. Soc.* **2019**, *141*, 14904–14915. [CrossRef] [PubMed]
116. Liang, X.T.; Chen, J.H.; Yang, Z. Asymmetric Total Synthesis of (−)-Spirochensilide A. *J. Am. Chem. Soc.* **2020**, *142*, 8116–8121. [CrossRef]
117. Xu, B.; Xun, W.; Su, S.; Zhai, H. Total Syntheses of (−)-Conidiogenone B, (−)-Conidiogenone, and (−)-Conidiogenol. *Angew. Chem. Int. Ed.* **2020**, *132*, 16617–16621.
118. Clark, J.S.; Xu, C. Total Synthesis of (−)-Nakadomarin A. *Angew. Chem. Int. Ed.* **2016**, *55*, 4332–4335. [CrossRef] [PubMed]
119. Lv, C.; Tu, Q.; Gong, J.; Hao, X.; Yang, Z. Asymmetric total synthesis of (−)-perforanoid A. *Tetrahedron* **2017**, *73*, 3612–3621. [CrossRef]
120. Cassayre, J.; Gagosz, F.; Zard, S.Z. A Short Synthesis of (±)-13-Deoxyserratine. *Angew. Chem. Int. Ed.* **2002**, *41*, 1783–1785. [CrossRef]
121. Nakayama, A.; Kogure, N.; Kitajima, M.; Takayama, H. Asymmetric Total Synthesis of a Pentacyclic Lycopodium Alkaloid: Huperzine-Q. *Angew. Chem. Int. Ed.* **2011**, *50*, 8025–8028. [CrossRef] [PubMed]
122. Itoh, N.; Iwata, T.; Sugihara, H.; Inagaki, F.; Mukai, C. Total Syntheses of (±)-Fawcettimine, (±)-Fawcettidine, (±)-Lycoflexine, and (±)-Lycoposerramine-Q. *Chem. Eur. J.* **2013**, *19*, 8665–8672. [CrossRef]
123. Williams, B.M.; Trauner, D. Expedient Synthesis of (+)-Lycopalhine A. *Angew. Chem. Int. Ed.* **2016**, *55*, 2191–2194. [CrossRef]
124. Yamakoshi, H.; Sawayama, Y.; Akahori, Y.; Kato, M.; Nakamura, S. Total Syntheses of (+)-Marrubiin and (−)-Marrulibacetal. *Org. Lett.* **2016**, *18*, 3430–3433. [CrossRef] [PubMed]
125. Sakagami, Y.; Kondo, N.; Sawayama, Y.; Yamakoshi, H.; Nakamura, S. Total syntheses of marrubiin and related labdane diterpene lactones. *Molecules* **2020**, *25*, 1610. [CrossRef] [PubMed]
126. Bose, S.; Yang, J.; Yu, Z.-X. Formal Synthesis of Gracilamine Using Rh(I)- Catalyzed [3 + 2 + 1] Cycloaddition of 1-Yne-Vinylcyclopropanes and CO. *J. Org. Chem.* **2016**, *81*, 6757–6765. [CrossRef] [PubMed]
127. Shi, Y.; Yang, B.; Cai, S.; Gao, S. Total Synthesis of Gracilamine. *Angew. Chem. Int. Ed.* **2014**, *53*, 9539–9543. [CrossRef] [PubMed]
128. Lopez-Perez, B.; Maestro, M.A.; Mourino, A. Total synthesis of 1α,25-dihydroxyvitamin D3 (calcitriol) through a Si-assisted allylic substitution. *Chem. Commun.* **2017**, *53*, 8144–8147. [CrossRef] [PubMed]
129. Salam, A.; Ray, S.; Zaid, M.A.; Kumar, D.; Khan, T. Total syntheses of several iridolactones and the putative structure of noriridoid scholarein A: An intramolecular Pauson–Khand reaction based one-stop synthetic solution. *Org. Biomol. Chem.* **2019**, *17*, 6831–6842. [CrossRef]
130. Kourav, M.S.; Kumar, V.; Kumar, D.; Khan, T. A De Novo Approach for the Stereoselective Synthesis of Cyclopenta[c]pyranone Scaffold Present in Iridoids: Formal Syntheses of Isoboonein, Iridomyrmecin and Isoiridomyrmecin. *ChemistrySelect* **2018**, *3*, 5566–5570. [CrossRef]
131. Hirasawa, Y.; Morita, H.; Shiro, M.; Kobayashi, J. Sieboldine A, a Novel Tetracyclic Alkaloid from Lycopodium sieboldii, Inhibiting Acetylcholinesterase. *Org. Lett.* **2003**, *5*, 3991–3993. [CrossRef] [PubMed]

132. Abd El-Gaber, M.K.; Yasuda, S.; Iida, E.; Mukai, C. Enantioselective Total Synthesis of (+)-Sieboldine A. *Org. Lett.* **2017**, *19*, 320–323. [CrossRef] [PubMed]
133. Tao, C.; Zhang, J.; Chen, X.; Wang, H.; Li, Y.; Cheng, B.; Zhai, H. Formal Synthesis of (+/−)-Aplykurodinone-1 through a Hetero-Pauson-Khand Cycloaddition Approach. *Org. Lett.* **2017**, *19*, 1056–1059. [CrossRef] [PubMed]
134. He, L.; Deng, L.-L.; Mu, S.-Z.; Sun, Q.-Y.; Hao, X.-J.; Zhang, Y.-H. Sinoraculine, the Precursor of the Novel Alkaloid Sinoracutine from Stephania cepharantha Hayata. *Helvetica Chimica Acta* **2012**, *95*, 1198–1201. [CrossRef]
135. Bao, G.-H.; Wang, X.-L.; Tang, X.-C.; Chiu, P.; Qin, G.-W. Sinoracutine, a novel skeletal alkaloid with cell-protective effects from *Sinomenium acutum*. *Tetrahedron Lett.* **2009**, *50*, 4375–4377. [CrossRef]
136. Volpin, G.; Veprek, N.A.; Bellan, A.B.; Trauner, D. Enantioselective Synthesis and Racemization of (−)-Sinoracutine. *Angew. Chem. Int. Ed.* **2017**, *56*, 897–901. [CrossRef] [PubMed]
137. Sennett, S.H.; Pompeni, S.A.; Wright, A.E. Diterpene Metabolites from Two Chemotypes of the Marine Sponge Myrmekioderma styx. *J. Nat. Prod.* **1992**, *55*, 1421–1429. [CrossRef]
138. Chang, Y.; Shi, L.; Huang, J.; Shi, L.; Zhang, Z.; Hao, H.D.; Gong, J.; Yang, Z. Stereoselective Total Synthesis of (+/−)-5-epi-Cyanthiwigin I via an Intramolecular Pauson-Khand Reaction as the Key Step. *Org. Lett.* **2018**, *20*, 2876–2879. [CrossRef]
139. Kaneko, H.; Takahashi, S.; Kogure, N.; Kitajima, M.; Takayama, H. Asymmetric Total Synthesis of Fawcettimine-Type Lycopodium Alkaloid, Lycopoclavamine-A. *J. Org. Chem.* **2019**, *84*, 5645–5654. [CrossRef]
140. Sadler, I.H.; Simpson, T.J. The determination by n.m.r. methods of the structure and stereochemistry of astellatol, a new and unusual sesterterpene. *J. Chem. Soc. Chem. Commun.* **1989**, *21*, 1602–1604. [CrossRef]
141. Zhao, N.; Xie, S.; Tian, P.; Tong, R.; Ning, C.; Xu, J. Asymmetric total synthesis of (+)-astellatol and (−)-astellatene. *Org. Chem. Front.* **2019**, *6*, 2014–2022. [CrossRef]
142. Chirkin, E.; Bouzidi, C.; Porée, F.H. Tungsten-Promoted Hetero-Pauson–Khand Cycloaddition: Application to the Total Synthesis of (−)-Allosecurinine. *Synthesis* **2019**, *51*, 2001–2006. [CrossRef]
143. Hugelshofer, C.L.; Palani, V.; Sarpong, R. Calyciphylline BType Alkaloids: Total Syntheses of (−)-Daphlongamine H and (−)-Isodaphlongamine H. *J. Am. Chem. Soc.* **2019**, *141*, 8431–8435. [CrossRef]
144. Winther, M.; Liu, H.; Sonntag, Y.; Olesen, C.; Le Maire, M.; Soehoel, H.; Olsen, C.E.; Christensen, S.B.; Nissen, P.; MÜller, J.V. Critical Roles of Hydrophobicity and Orientation of Side Chains for Inactivation of Sarcoplasmic Reticulum Ca^{2+}-ATPase with Thapsigargin and Thapsigargin Analogs. *J. Biol. Chem.* **2010**, *285*, 28883–28892. [CrossRef]
145. Skytte, D.M.; MÜller, J.V.; Liu, H.; Nielsen, H.Ø.; Svenningsen, L.E.; Jensen, C.M.; Olsen, C.E.; Christensen, S.B. Elucidation of the Topography of the Thapsigargin Binding Site in the Sarco-endoplasmic Calcium ATPase. *Bioorg. Med. Chem.* **2010**, *18*, 5634–5646. [CrossRef] [PubMed]
146. Denmeade, S.R.; Mhaka, A.M.; Rosen, D.M.; Brennen, W.N.; Dalrymple, S.; Dach, I.; Olesen, C.; Gurel, B.; DeMarzo, A.M.; Wilding, G.; et al. Engineering a Prostate-Specific Membrane Antigen-Activated Tumor Endothelial Cell Prodrug for Cancer Therapy. *Sci. Transl. Med.* **2012**, *4*, 140ra86. [CrossRef] [PubMed]
147. Lynch, J.K.; Hutchison, J.J.; Fu, X.; Kunnen, K. Methods of Making Cancer Compositions. WO 2014145035 A1, 18 September 2014.
148. Zimmermann, T.; Christensen, S.B.; Franzyk, H. Preparation of Enzyme-Activated Thapsigargin Prodrugs by Solid-Phase Synthesis. *Molecules* **2018**, *23*, 1463. [CrossRef]
149. Sanogo, Y.; Othman, R.B.; Dhambri, S.; Selkti, M.; Jeuken, A.; Prunet, J.; Ferezou, J.P.; Ardisson, J.; Lannou, M.I.; Sorin, G. Ti(II) and Rh(I) Complexes as Reagents toward a Thapsigargin Core. *J. Org. Chem.* **2019**, *84*, 5821–5830. [CrossRef]
150. Tian, H.Y.; Ruan, L.J.; Yu, T.; Zheng, Q.F.; Chen, N.H.; Wu, R.B.; Zhang, X.Q.; Wang, L.; Jiang, R.W.; Ye, W.C. Bufospirostenin A and Bufogargarizin C, Steroids with Rearranged Skeletons from the Toad Bufo Bufo Gargarizans. *J. Nat. Prod.* **2017**, *80*, 1182–1186. [CrossRef]
151. Cheng, M.J.; Zhong, L.P.; Gu, C.C.; Zhu, X.J.; Chen, B.; Liu, J.S.; Wang, L.; Ye, W.C.; Li, C.C. Asymmetric Total Synthesis of Bufospirostenin A. *J. Am. Chem. Soc.* **2020**, *142*, 12602–12607. [CrossRef]
152. Closser, K.D.; Quintal, M.M.; Shea, K.M. The Scope and Limitations of Intramolecular Nicholas and Pauson–Khand Reactions for the Synthesis of Tricyclic Oxygen- and Nitrogen-Containing Heterocycles. *J. Org. Chem.* **2009**, *74*, 3680–3688. [CrossRef] [PubMed]

153. Comer, E.; Rohan, E.; Deng, L.; Porco, J.A. An Approach to Skeletal Diversity Using Functional Group Pairing of Multifunctional Scaffolds. *Org. Lett.* **2007**, *9*, 2123–2126. [CrossRef]
154. Nie, F.; Kunciw, D.L.; Wilcke, D.; Stokes, J.E.; Galloway, W.R.; Bartlett, S.; Sore, H.F.; Spring, D.R. A Multidimensional Diversity-Oriented Synthesis Strategy for Structurally Diverse and Complex Macrocycles. *Angew. Chem. Int. Ed.* **2016**, *55*, 11139–11143. [CrossRef] [PubMed]

Publisher's Note: MDPI stays neutral with regard to jurisdictional claims in published maps and institutional affiliations.

© 2020 by the authors. Licensee MDPI, Basel, Switzerland. This article is an open access article distributed under the terms and conditions of the Creative Commons Attribution (CC BY) license (http://creativecommons.org/licenses/by/4.0/).

Review

Radical Carbonylative Synthesis of Heterocycles by Visible Light Photoredox Catalysis

Xiao-Qiang Hu [1,*], Zi-Kui Liu [1] and Wen-Jing Xiao [2,*]

[1] Key Laboratory of Catalysis and Energy Materials Chemistry of Ministry of Education & Hubei Key Laboratory of Catalysis and Materials Science, School of Chemistry and Materials Science, South-Central University for Nationalities, Wuhan 430074, China; liuzikui2020@163.com
[2] Key Laboratory of Pesticide & Chemical Biology Ministry of Education, College of Chemistry, Central China Normal University (CCNU), Wuhan 430079, China
* Correspondence: huxiaoqiang@mail.scuec.edu.cn (X.-Q.H.); wxiao@mail.ccnu.edu.cn (W.-J.X.)

Received: 1 August 2020; Accepted: 11 September 2020; Published: 14 September 2020

Abstract: Visible light photocatalytic radical carbonylation has been established as a robust tool for the efficient synthesis of carbonyl-containing compounds. Acyl radicals serve as the key intermediates in these useful transformations and can be generated from the addition of alkyl or aryl radicals to carbon monoxide (CO) or various acyl radical precursors such as aldehydes, carboxylic acids, anhydrides, acyl chlorides or α-keto acids. In this review, we aim to summarize the impact of visible light-induced acyl radical carbonylation reactions on the synthesis of oxygen and nitrogen heterocycles. The discussion is mainly categorized based on different types of acyl radical precursors.

Keywords: radical carbonylation; acyl radical; visible light photocatalysis; heterocycles

1. Introduction

Carbonyl-containing compounds, such as ketones, esters and amides, widely exist in numerous biologically important natural products, functional materials as well as pharmaceuticals [1–5]. The development of efficient methods toward synthesis of these substantial compounds has been intensively pursued by synthetic chemists over the past decades. Radical carbonylation promoted by transition metals (Pd, Mn, Co, Ni, Ru, Rh etc.), external oxidants or thermal initiation has been well-established for the preparation of carbonyl compounds with high efficiency [6–12]. Additionally, light-induced radical carbonylation reaction affords an alternative platform for the assembly of carbonyl motifs [13–17]. The pioneering examples are typically mercury and polyoxotungstate-photosensitized alkane carbonylation, metal-carbonyl catalyzed radical carbonylation and radical/Pd-combined carbonylation. Despite these impressive advances, the requirement of high-energy UV irradiation, poor selectivity or low efficiency profoundly limits their broad applications in practical synthesis.

Visible light photoredox catalysis has emerged as one of the most important techniques in radical reactions by employing an abundant and endlessly renewable solar energy as a driving force [18–30]. Under photocatalytic conditions, various radical species can be formed in a mild and controllable fashion, which enables the precise synthesis of high value-added products. In this context, acyl radicals are commonly generated by the addition of alkyl or aryl radicals to carbon monoxide (CO) or from single electron transfer (SET) conversion of aldehydes, carboxylic acids, anhydrides, acyl chlorides or α-keto acids (Scheme 1) [31]. In this review, we mainly summarize recent advances in the field of visible light-induced radical carbonylative synthesis of oxygen and nitrogen heterocycles with an emphasis on the catalytic system, scope and reaction mechanism. The principal achievements are partitioned into the sections organized based on different acyl radical precursors. Transition metal-catalyzed radical carbonylation and atom transfer radical carbonylation have been comprehensively reviewed before and will not be discussed here [32,33].

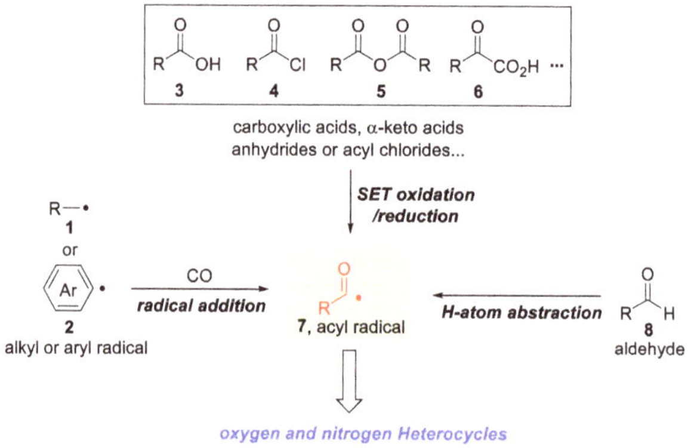

Scheme 1. Acyl radical carbonylation reactions in the synthesis of heterocycles.

2. Carbon Monoxide (CO)-Mediated Radical Carbonylation

Carbon monoxide (CO), as a cheap and readily available carbonyl source, has been extensively used in radical carbonylations [34,35]. In 2014, Xiao and colleagues reported a visible light-induced photocatalytic alkoxycarboxylation of aryldiazonium salts with alcohols under the CO pressure of 80 atm (Scheme 2) [36]. A range of structurally diverse benzoates was obtained in moderated to good yields. Notably, natural chiral alcohols such as N-benzoyl L-(+)-prolinol, (-)-menthol and methyl D-(-)-mandelate were compatible with this carbonylation reaction. When *ortho*-allyl- or *ortho*-propargyl-substituted benzenediazoniums were used, the resultant dihydrobenzofuran **11d** and benzofuran **11e** were formed in 63% and 58% yields, respectively. Almost at the same time, the group of Wangelin achieved the same reaction using eosin Y as a cheap photosensitizer [37]. Importantly, a lower pressure of CO (50 atm) was sufficient in this reaction. These two findings opened new avenues for the development of radical carboxylation reactions.

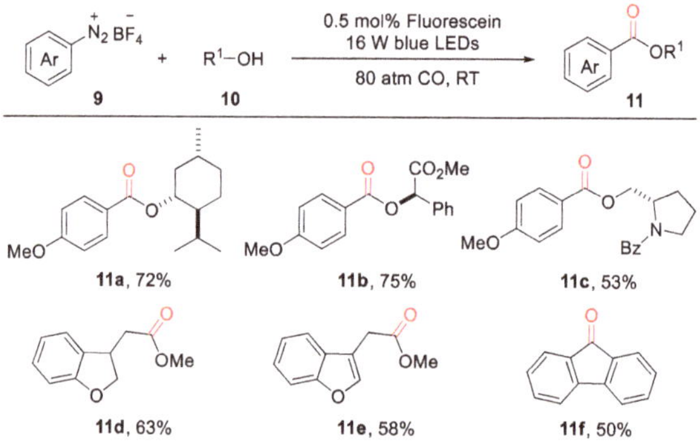

Scheme 2. Alkoxycarboxylation of aryldiazonium salts with alcohols.

It is believed that a visible light-induced radical carbonylation pathway would be involved in this reaction, as accounted for by control experiments and density functional theory (DFT) calculations (Scheme 3). Upon irradiation with blue LEDs, the photocatalyst PC initially undergoes a photoexcitation process to give a long-lived excited species PC*. Then, a single-electron reduction of aryldiazonium salt **9** by PC* produces the intended aryl radical intermediate **9-A**. Thereafter, radical **9-A** is rapidly trapped by CO to generate the key acyl radical **9-B**, which can be easily oxidized by PC$^+$ to achieve the benzylidyneoxonium **9-C**, thus completing the photocatalytic cycle. Subsequently, the electronic trapping of **9-C** by alcohols results in the benzoate products.

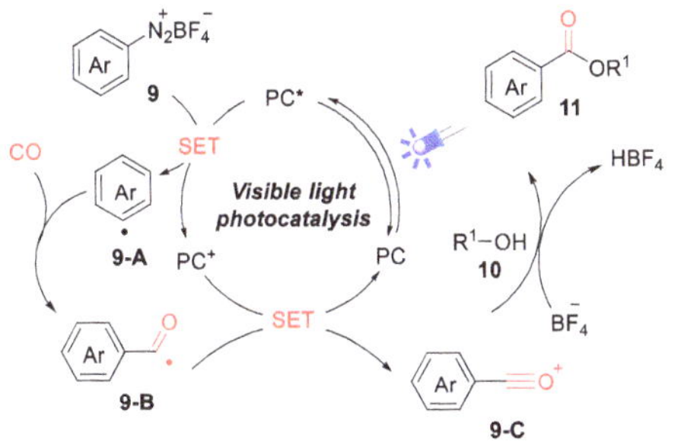

Scheme 3. A plausible mechanism for the alkoxycarboxylation of aryldiazonium salts.

Later, Gu et al. used (hetero) arenes as electrophiles to trap the reactive benzylidyneoxonium intermediates, delivering aryl ketones in moderate yields [38]. Generally, the electron-rich arenes gave better results over electron-deficient substrates, and nitrobenzene failed to give the expected product in the current system. In 2016, Li et al. further extended this strategy to the carbonylation of indoles for the preparation of various indol-3-yl aryl ketones [39]. In addition, by employing readily available arylsulfonyl chlorides as a robust source of aryl radicals, Liang, Li and co-workers achieved the same reaction under a 80 atm CO atmosphere with irradiation by green LEDs [40].

In 2018, a visible light-induced annulative carbonylation of alkenyl-tethered aryldiazonium salts was developed by Polyzos et al. in continuous-flow (Scheme 4) [41]. Under a moderate CO pressure (25 atm), this reaction proceeded smoothly to afford a range of 3-acetate substituted 2,3-dihydrobenzofurans **14** in satisfactory yields with excellent chemo- and regio-selectivity. Moreover, the generality of this transformation can be further extended to the reaction of unsaturated *ortho*-tethered aryldiazonium salts. When 1-butenyloxy- and propargyloxy-substituted aryldiazonium salts were subjected to the optimal conditions, the acetate-functionalized chromane **14f** and benzofuran **14e** were successfully obtained in 72% and 57% yields, respectively. It should be noted that the current continuous-flow protocol drastically shortens the reaction time to 200 s, which enables a preparative scale-up reaction in high efficiency.

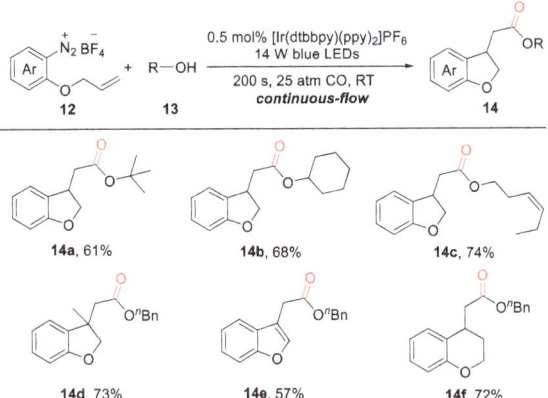

Scheme 4. Annulative carbonylation of alkenyltethered aryldiazonium salts.

In 2015, Xiao and Lu et al. reported an elegant decarboxylative/carbonylative alkynylation of carboxylic acids (Scheme 5a) [42]. Remarkable features of this transformation include mild conditions and high efficiency. The synthetic utility of this methodology was demonstrated by the success of a rapid solar light-driven reaction and 5.0 mmol gram-scale reaction. Unlike the well-established Heck reactions of alkenyl and aryl substrates, the alkyl-Heck reaction is still problematic due to the inherent difficulty in palladium-mediated oxidative addition of electrophilic alkyl reagents and competitive β-hydrogen elimination of the generated alkyl–palladium species. Very recently, Xiao and Lu et al. further disclosed a novel deaminative alkyl-Heck-type reaction by replacing palladium-catalyzed two-electron pathway with a photocatalytic single-electron activation [43]. Katritzky salts acted as the alkyl radical sources in this reaction, which can be easily synthesized from structurally diverse alkyl amines [44–46]. The choice of bases is critical in this reaction and 1,4-diazabicyclo[2.2.2]octane turned out to be the optimal base. Moreover, under a CO pressure of 80 atm, a photostimulated deaminative/carbonylative Heck-type procedure was developed for the construction of α,β-unsaturated ketones in moderate to excellent yields (Scheme 5b). This radical-mediated alkyl-Heck-type protocol provides a new approach to the libraries of alkenes, complementing the current late-stage of palladium-catalyzed Heck reactions. As illustrated in Scheme 5b, under irradiation with blue LEDs, the excited state Ir(III)* is quickly quenched by Katritzky salt **18** to give a reactive alkyl radical **18-A** along with the formation of Ir(IV)-catalyst. The subsequent reaction of alkyl radical **18-A** with CO produces an acyl radical **18-B**, which then reacts with an alkene to give radical species **19-A**. Finally, the SET oxidation of **19-A** by Ir(IV)-catalyst affords intermediate **19-B**, which undergoes a deprotonation process to give the final enone product.

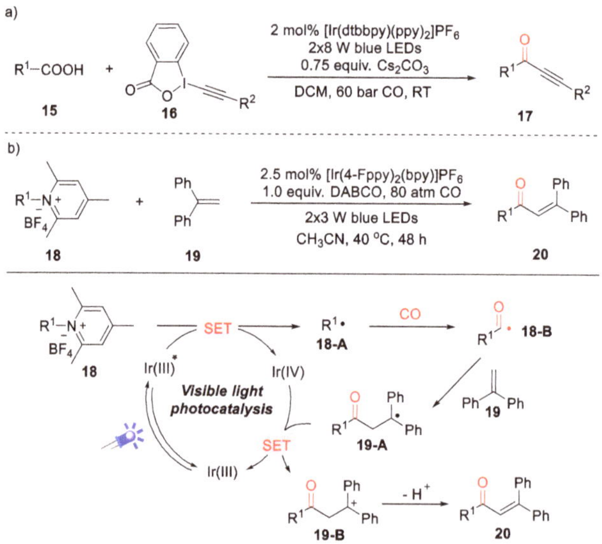

Scheme 5. (**a**) Decarboxylative/carbonylative alkynylation of carboxylic acids; (**b**) Photostimulated deaminative alkyl-Heck-type reaction.

A novel photocatalytic aminocarbonylation of cycloketone oxime esters with amines has been discovered by Xiao and Chen (Scheme 6) [47]. The reactive Cu(I) complex derived from CuCl catalyst and N, N, N-tridentate ligand worked as both visible light photocatalyst and carbonylation catalyst in this reaction. Control experiments suggested the necessity of a copper catalyst, ligand and visible light irradiation. Under the standard conditions, a variety of (hetero)aryl amines and alkyl amines reacted with cycloketone oxime esters, yielding the structurally different aminocarbonylation products in moderate to good yields (Scheme 6a). Importantly, the pharmaceutical agent (±)-mexiletine can be converted in this reaction, highlighting the synthetic potential of this methodology. This reaction provides a direct access to diverse cyanoalkylated amides at ambient temperature.

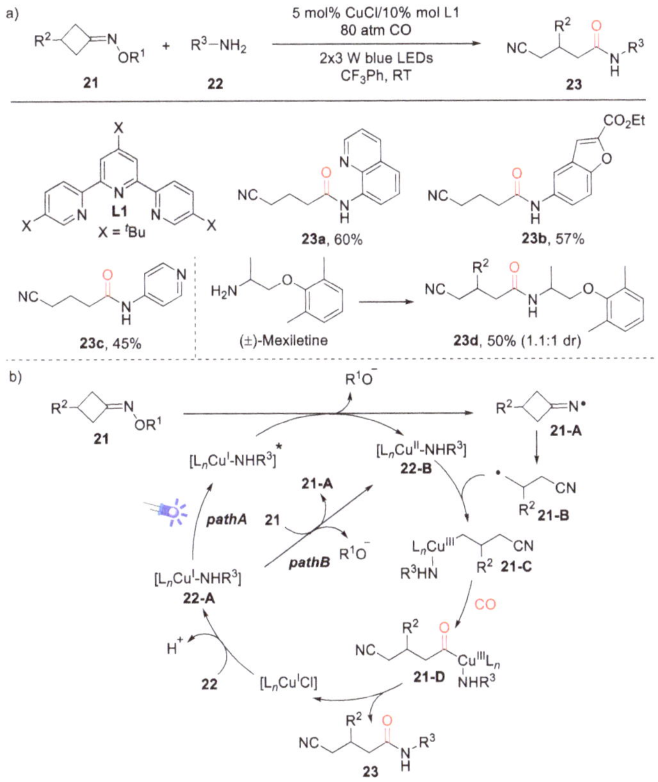

Scheme 6. (**a**) Aminocarbonylation of oxime esters; (**b**) Proposed mechanism.

According to the mechanistic studies and literature precedents, two reaction pathways have been proposed by authors (Scheme 6b). Under blue-light irradiation, a single-electron reduction of cycloketone oxime ester **21** by the photoexcited state of $L_nCu(I)$–NHPh complex **22-A*** (path A) is observed, delivering an iminyl radical intermediate **21-A**. In addition, the cycloketone oxime ester **21** can be also reduced by the ground state of $Cu(I)/L_n$ catalyst **22-A** due to its redox reactivity (path B). Then, a fast β–C–C bond cleavage of **21-A** leads to the cyanoalkyl radical intermediate **21-B**, which can react with Cu(II)-catalyst to generate a high-valent Cu(III) complex **21-C**. The insertion of a CO molecule into Cu(III)–C bond affords the intermediate **21-D**. Finally, the reductive elimination of **21-D** gives rise to the amide product **23** with the regeneration of Cu(I)-catalyst for the next catalytic cycle.

Very recently, Arndtsen et al. developed an elegant carbonylative coupling reaction of aryl or alkyl halides with some challenging nucleophiles [48]. The active catalysts of this transformation were believed to be the photoexcited state of Pd(0) and Pd(II) species, which can promote the oxidative addition and reductive elimination steps with low energy barriers (Scheme 7a). It was found that the addition of a visible light photocatalyst was not necessary for this reaction. However, in the absence of blue-light irradiation, palladium catalyst or phosphine ligand (DPEphos), no reaction has been observed. A range of nucleophiles such as sterically hindered secondary amines, tertiary alcohols, substituted anilines and even weakly nucleophilic N-heterocycles can be coupled at ambient temperature, producing various important esters, amides and ketones.

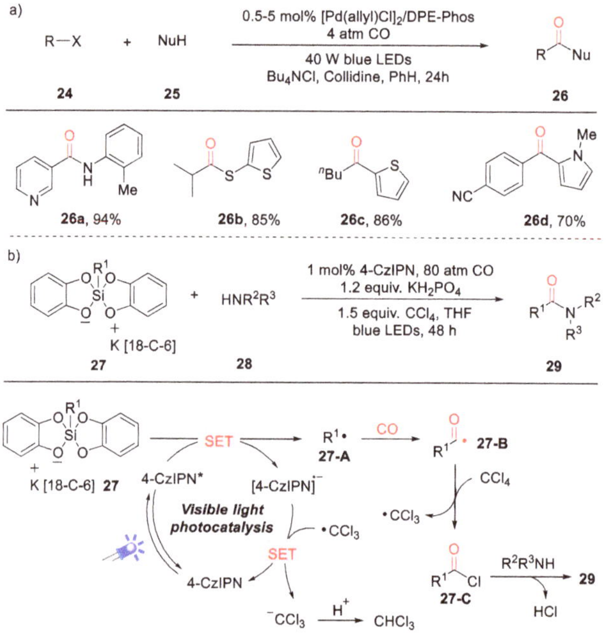

Scheme 7. (**a**) Radical carbonylation of halides; (**b**) Radical carbonylation of organosilicates.

By using organosilicates as alkyl radical precursors, Fensterbank, Ryu, Ollivier and Fukuyama et al. demonstrated a radical carbonylation of various amines for the construction of aliphatic amides under 80 atm pressure of CO (Scheme 7b) [49]. In this reaction, CCl$_4$ acted as a Cl-atom donor to react with acyl radicals for the in situ formation of acyl chlorides, which then reacted with amines to produce amide products. CBrCl$_3$ was also suitable for this process, albeit with a lower yield. The pressure of CO has a pronounced effect on this reaction. When a lower CO pressure of 40 atm was used, the chemical yield was significantly decreased. This strategy can be further applied in an intramolecular carbonylation reaction, leading to pyrrolidinone as the sole product. A plausible mechanism is described in Scheme 7b. Under the irradiation by blue LEDs, the excitation of photocatalyst 4-CzIPN delivers 4-CzIPN*, which initially reacts with organosilicate **27** to give an alkyl radical **27-A** via an oxidative Si–C bond cleavage. The addition of radical **27-A** to CO results in an acyl radical **27-B**, which then abstracts a chlorine atom from CCl$_4$ to generate acyl chloride **27-C** and trichloromethyl radical. Acyl chloride **27-C** is rapidly trapped by amines to give the amide products. The in situ formed trichloromethyl radical can regenerate the photocatalyst 4-CzIPN through a single electron-transfer process.

Transition metal-catalyzed oxidative carbonylation provides an effective and general platform for the construction of carbonyl compounds [6]. In this type of reactions, a stoichiometric amount of external oxidants is often required for the recyclization of the active Pd(II)-catalyst. Taking the advantage of the fact that O$_2$ is a green and powerful oxidant, the group of Lei achieved a novel O$_2$-mediated oxidative carbonylation of enamides by merging photoredox catalysis and palladium catalysis under a low CO pressure (Scheme 8a) [50]. It should be noted that the usage of 8 mol% xantphos can chiefly improve yields of the desired products. ^{31}P NMR experiments suggested that the phosphine ligand would be firstly converted into its oxidized species, thus facilitating the oxidative carbonylation process. This dual catalytic system affords an environmentally benign access to 1,3-oxazin-6-ones.

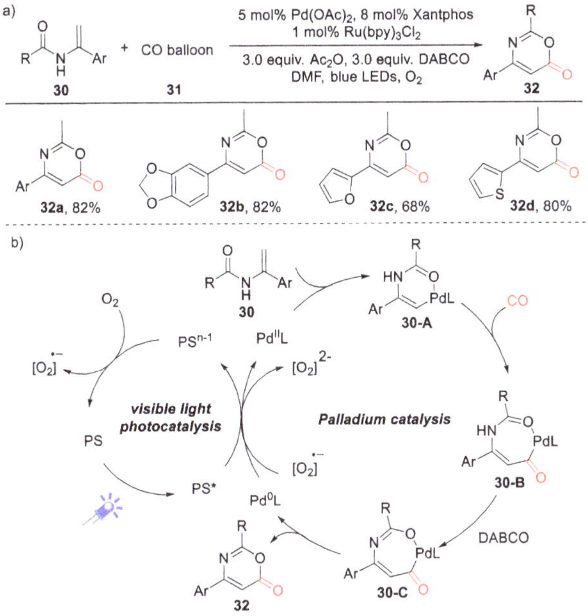

Scheme 8. (a) Oxidative carbonylation of enamides; (b) Proposed mechanism.

As described in Scheme 8b, the vinylpalladium intermediate **30-A** is firstly generated from enamide via Pd(OAc)$_2$-promoted C(sp^2)–H bond activation of enamides. Then, the coordination and insertion of a CO molecule into intermediate **30-A** produce acylpalladium complex **30-B**, which can be transferred into **30-C** in the presence of 1,4-diazabicyclo[2.2.2]octane (DABCO). Reductive elimination of **30-C** delivers the final carbonylation product **32** with the formation of Pd(0)-catalyst. The Pd(0)-catalyst can be further oxidized by the photoexcited state of photocatalyst or superoxide anion to the active Pd(II)-catalyst, thus completing the palladium catalytic cycle.

3. Aldehyde-Mediated Radical Carbonylation

Aldehydes have been served as versatile building blocks in organic synthesis for a long time. In the presence of hydrogen atom transfer (HAT), reagents such as persulfates, *tert*-butyl hydroperoxide (TBHP), quinuclidine and Eosin Y, aldehydes can be easily converted into acyl radical species through a rapid HAT [31,51]. In this section, we mainly discuss the acyl radical reactions of aldehydes for the assembly of oxygen and nitrogen heterocycles. In 2014, Zeng and Xie et al. reported an interesting benzaldehyde-mediated Minisci reaction for the regiospecific acylation of biologically important phenanthridines (Scheme 9) [52]. In this reaction, 2.0 equivalent of (NH$_4$)$_2$S$_2$O$_8$ was used as the HAT reagent as well as the external oxidant. This catalytic system is especially well-adapted for aromatic aldehydes, giving 6-acylated phenanthridines in moderate yields.

As an attractive alternative to persulfates, the cheap and readily available TBHP can be also used as a good HAT reagent [53]. In 2015, Wang and co-workers independently published an acyl radical-mediated cascade reaction of benzaldehydes and styrenes for the synthesis of α, β-epoxy ketones (Scheme 10) [54]. TBHP was identified to be superior over other traditional oxidants such as K$_2$S$_2$O$_8$, benzoyl peroxide (BPO) and (t-BuO)$_2$. Various styrenes bearing electron-donating and -withdrawing groups performed well, providing α, β-epoxy ketones in generally good yields. Notably, both 1,1-disubstituted alkenes and pentafluorinated alkenes were compatible with this catalytic system. Significantly, this reaction can be easily scaled up for a gram-scale ketone synthesis.

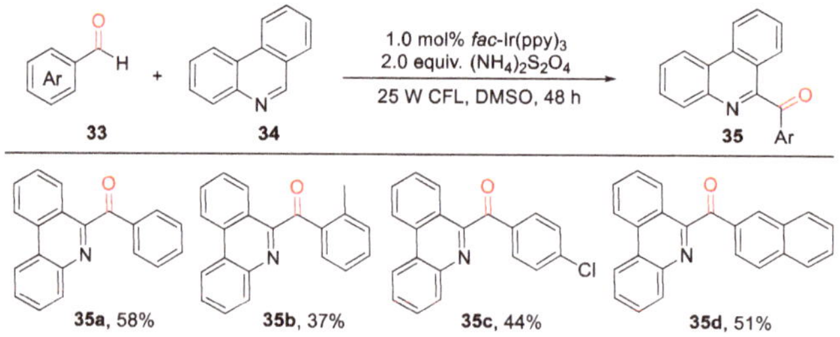

Scheme 9. Benzaldehyde-mediated regiospecific acylation of phenanthridines.

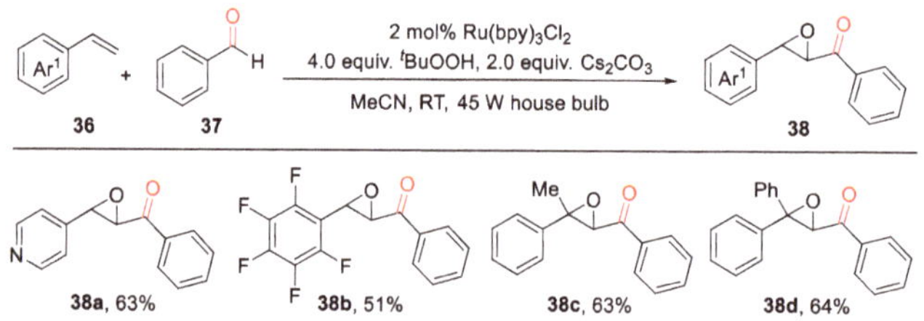

Scheme 10. Acyl radical-mediated cascade reaction of benzaldehydes and styrenes.

Mechanistic studies by radical trapping experiments and analysis of the reaction mixture by high-resolution mass spectroscopy (HRMS) suggested a plausible acyl radical pathway (Scheme 11). At first, an oxidative quenching of the photoexcited Ru^{2+*} complex by TBHP gives a *tert*-butoxy radical, which subsequently abstracts a H-atom from benzaldehyde to form the key acyl radical **37-A**. Secondly, the selective addition of radical **37-A** to alkene **36** results in a new C-based radical intermediate **36-A**. At the same time, the oxidation of *tert*-butyl peroxide anion by oxidizing Ru^{3+} complex affords *tert*-butyl peroxide radical and regenerates the ground state of Ru^{2+} catalyst. The rapid radical coupling of t-BuOO• with **36-A** delivers β-peroxy ketone **36-B**, which can be detected by HRMS analysis. Finally, under basic conditions, **36-B** undergoes a cyclization process to generate the desired α, β-epoxy ketone product with the elimination of ^{t}BuOH.

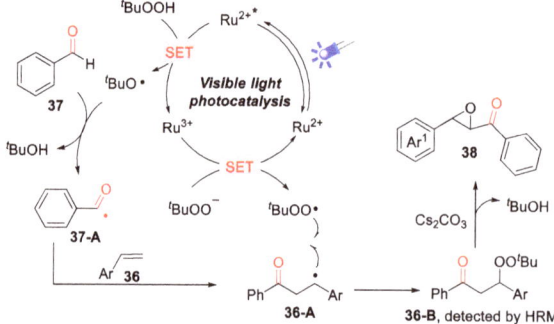

Scheme 11. Proposed pathway for the radical cascade reaction of benzaldehydes and styrenes.

The group of Salles Jr achieved the same reaction in water with the use of persulfate $K_2S_2O_8$ as the oxidant and methylene blue as the organophotoredox catalyst (Scheme 12) [55]. It is noteworthy that O_2 in water plays an important role in this transformation. Lower yields were observed when reactions were conducted in the degassed water or under N_2 atmosphere. Two different types of reactions have been developed, employing only one set of reagents. When styrenes were used as radical acceptors, a visible light-induced epoxyacylation reaction was accomplished. Both aromatic and aliphatic aldehydes were smoothly converted into various epoxyketones in good yields. Interestingly, when nonconjugated olefins were used, the reaction underwent a direct hydroacylation process to form long-chain ketones. In this hydroacylation reaction, only aromatic aldehydes were tolerated. In 2020, the group of Kokotos developed an acyl radical-mediated hydroacylation reaction of alkenes from simple aldehydes under metal-free conditions [56]. Using 4-acyl-1,4-dihydropyridines as acyl radical sources, Xia et al. reported an interesting visible light-induced hydroacylation of alkenes under photocatalyst-free conditions. Additionally, in the presence of a Ni(II)-catalyst, the diacylation reaction was developed [57]. Moreover, Melchiorre et al. achieved an asymmetric acylation reaction of enals for the stereocontrolled construction of 1,4-dicarbonyl products [58].

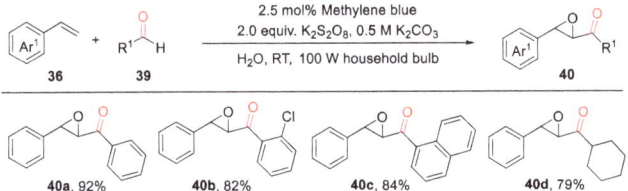

Scheme 12. Visible light-induced acyl radical epoxyacylation of olefins.

Using the same strategy, in 2017, Hong's group reported an acyl radical-mediated intramolecular epoxyacylation reaction catalyzed by Ru(bpy)$_3$Cl$_2$ (Scheme 13a) [59]. The investigation of optimum reaction conditions revealed that TBHP was the best HAT reagent. Under optimal conditions, various spiroepoxy chroman-4-one scaffolds and spiroepoxy enaminones could be constructed in moderate yields. The utility was demonstrated by mild conditions, simple operation and broad substrate scope. In addition, the applicability of this transformation can be further extended to the epoxyacylation of benzylic alcohols, albeit with relatively lower yields (Scheme 13b). In the tandem reaction of benzylic alcohols, 8.0 equivalent of TBHP was required for the in situ formation of benzaldehyde intermediates. This transformation is initiated by the generation of an aldehyde that undergoes a sequential H-atom transfer, intramolecular radical cyclization and epoxidation process to produce the final products.

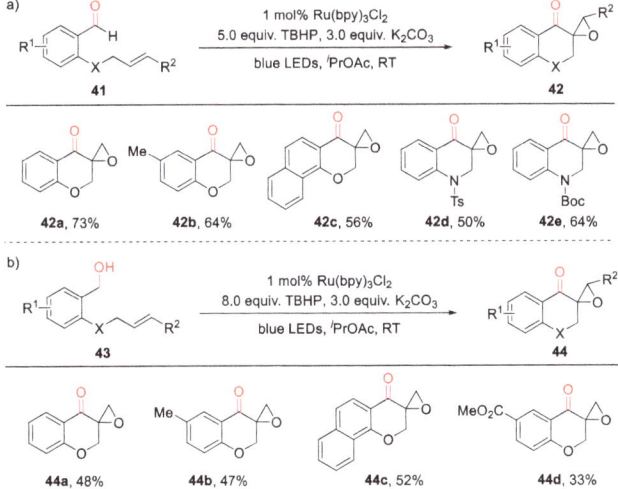

Scheme 13. (**a**) Intramolecular radical cylization/epoxyacylation of olefins; (**b**) Epoxyacylation of benzylic alcohols.

In 2018, Itoh and co-workers described an acyl radical-mediated addition/cyclization cascade reaction of ynoates with simple aldehydes by employing benzoyl peroxide (BPO) as the external oxidant and 2-*tert*-butylanthraquinone (2-^{t}Bu-AQN) as the photocatalyst (Scheme 14a) [60]. More than 20 coumarin derivatives can be synthesized in high efficiency, along with excellent regioselectivity. Notably, it was found that many of coumarin products have good antiproliferative activities against prostate cancer cells. The group of Klussmann developed a facile acyl radical difunctionalization of styrenes with the use of indoles and benzotriazole as nucleophiles (Scheme 14b) [61]. Very recently, a visible light photocatalytic deuteration of formyl groups was achieved by Wang et al. via the acyl radical-mediated H/D exchange strategy [62].

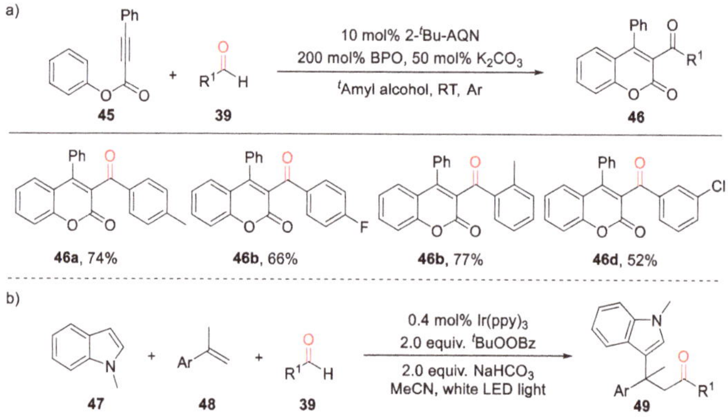

Scheme 14. (a) Acyl radical cascade reactions of ynoates; (b) Acyl radical difunctionalization of styrenes.

In addition, acyl radical-mediated alkynylation, arylation, vinylation and alkylation have been well-developed over the past 10 years. This has been reviewed by Ngai [31] and will not be discussed here.

4. Carboxylic Acids and Their Derivative-Mediated Radical Carbonylation

Carboxylic acids and their derivatives are promising chemical feedstocks in organic synthesis, and can be easily obtained in great structural diversity both from natural sources and some well-established methods [63–65]. Over the past decades, transition-metal catalyzed two-electron decarboxylative conversions of carboxylic acids have been well investigated toward a wide variety of valuable compounds. Recently, a visible light-driven single-electron transfer strategy has provided an important and new platform for the functionalizations of carboxylic acids and their derivatives. In this context, carboxylic acids and their derivatives can be selectively converted to acyl radicals for the preparation of carbonyl-containing compounds and various heterocycles [66].

The direct conversion of carboxylate groups into acyl radicals is relatively challenging due to their high bond strength (102 kcal/mol). In 2018, Zhu, Xie and co-workers developed a convenient deoxygenative activation of carboxylic acids for the hydroacylation of alkenes by using triphenylphosphine (Ph$_3$P) as the oxygen transfer reagent (Scheme 15a) [67]. Taking advantage of the strong P–O affinity between the Ph$_3$P radical cation and carboxylate anion, the homolytic cleavage of C–O bonds can be achieved to generate acyl radicals under mild photocatalytic conditions. Under the standard conditions, various aromatic acids were well-tolerated, while aliphatic acids proved to be unsuitable for this reaction. In addition, the intramolecular radical cyclization reactions were also investigated to synthesize cyclophane-braced macrocycloketones. Moreover, this methodology can be applied for the 3-step concise construction of the drug zolpidem.

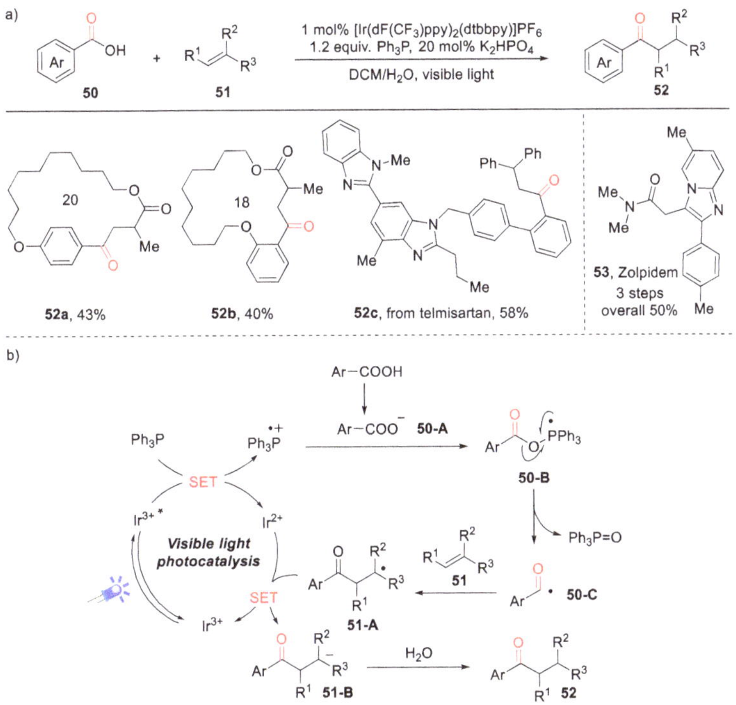

Scheme 15. (a) Photoinduced deoxygenative activation of carboxylic acids; (b) Proposed mechanism.

The reactions were completely inhibited by radical inhibitors such as 2,6-di-*tert*-butyl-*p*-cresol (BHT) and 2,2,6,6-tetramethyl-1-piperidyloxy (TEMPO). To further verify the origin of O-atom in the byproduct $Ph_3P=O$, ^{18}O-labeling experiments were investigated, suggesting that the O-atom of $Ph_3P=O$ would come from benzoic acids rather than from water. Taking together, a plausible acyl radical mechanism is illustrated in Scheme 15b. Under blue-light irradiation, Ph_3P is initially oxidized by the excited state of photocatalyst *Ir^{III} to generate Ph_3P radical cation. The rapid combination of Ph_3P radical cation with carboxylate anion **50-A** forms the phosphoranyl radical intermediate **50-B**. This intermediate undergoes a fast β-scission fragmentation and delivers acyl radical **50-C**, which then reacts with alkene to give C-radical **51-A**. A SET reduction of C-radical **51-A** by Ir^{II}-catalyst results in an anion intermediate **51-B**, followed by a protonation process to produce the ketone product **52**.

Almost simultaneously, Doyle and Rovis et al. applied this synthetic strategy in the general deoxygenative reduction of carboxylic acids via photoredox catalysis (Scheme 16a) [68]. Remarkably, both aromatic and aliphatic acids could be selectively converted to aldehydes by careful modification of the phosphine reagent. Ph_3P has been identified as a good oxygen transfer reagent for the deoxygenation of aromatic acids, while it was ineffective for the aliphatic acids. It is believed that a more electron-rich phosphoranyl radical species can be formed from Ph_3P and aliphatic carboxylic acid, which may undergo a rapid oxidation process to generate a reactive phosphonium intermediate for the acyl transfer reactions. As a result, an electron-deficient Ph_2POEt was selected as an optimal reductant for the reduction of aliphatic acids. Particularly, this method also enables conversions of carboxylic acids into cyclic ketones and lactones via intramolecular cyclization reactions of acyl radicals. More importantly, this strategy can be further applied in deoxygenative reduction of benzylic alcohols.

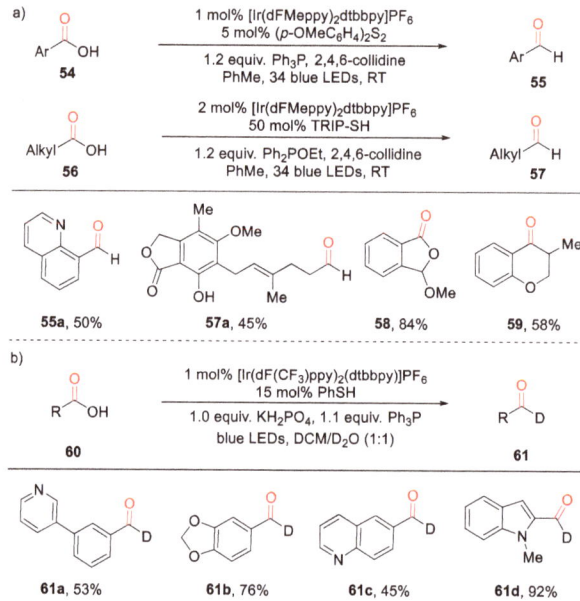

Scheme 16. (a) Photocatalytic deoxygenative reduction of carboxylic acids; (b) Deoxygenative deuteration of carboxylic acids.

After that, Xie et al. reported a novel deoxygenative deuteration of carboxylic acids by employing inexpensive D$_2$O as the deuterium source (Scheme 16b) [69]. The usage of a thiol catalyst significantly affects this transformation, which may act as a HAT catalyst to tune the equilibrium with D$_2$O, thus facilitating the formation of deuterated aldehydes. Both aromatic and aliphatic acids were conveniently reduced, delivering deuterated aldehydes in moderate yields with high levels of D-incorporation. The robustness of this methodology was demonstrated by the deuteration of biologically important pharmaceuticals and natural products as well as the downstream construction of D-labeled N-containing heterocycles. In a subsequent investigation, in 2019, Zhu, Xie and co-workers disclosed an interesting deoxygenative arylation of aryl carboxylic acids for the preparation of a broad range of unsymmetrical ketones via a visible light-induced 1,5-aryl migration process [70].

By using the same concept, Chu and Sun et al. achieved an elegant acyl radical-mediated intramolecular cyclization of aromatic acids for the preparation of various dibenzocycloketones (Scheme 17) [71]. Methylene blue was the optimal photocatalyst and O$_2$ as a green oxidant. A range of dibenzocycloketone derivatives were produced in good yields under metal-free conditions. Shortly thereafter, an intermolecular hydroacylation of styrenes was discovered by Doyle's group, affording various dialkyl ketone products [72]. The key to the success of this protocol was the rational selection of a phosphine reagent for the generation of acyl radicals. The electron-rich PMe$_2$Ph with a low oxidation potential was the best choice, which could outcompete the side reactions of alkene substrates. In 2020, Wang and co-workers disclosed a convenient deoxygenation/defluorination cascade reaction for the assembly of γ, γ-difluoroallylic ketones from α-trifluoromethyl alkenes and aryl carboxylic acids. Phenylacetic acids were also tolerated well, while other simple aliphatic carboxylic acids were not compatible with this reaction system [73].

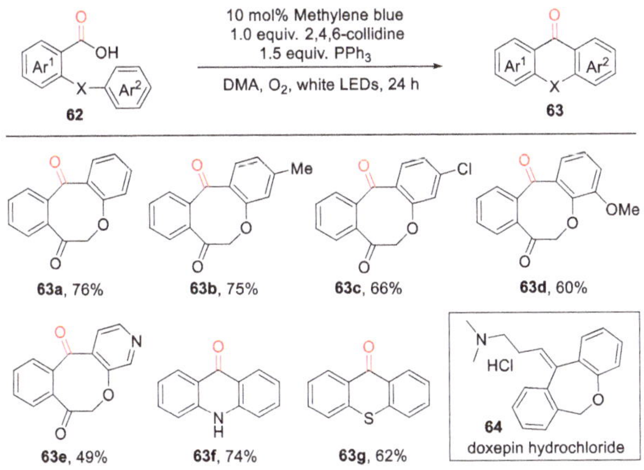

Scheme 17. Acyl radical-mediated intramolecular cyclization of aromatic acids.

The in situ generated anhydride from carboxylic acid could be engaged as an effective acyl radical precursor. In 2015, the group of Wallentin developed an interesting acylarylation cascade reaction of aromatic carboxylic acids and methacrylamides for the preparation of structurally diverse 3, 3-disubstituted 2-oxindoles (Scheme 18a) [74]. The transient anhydrides were believed to be formed from carboxylic acids and dimethyl dicarbonate (DMDC). The choice of visible light photocatalyst plays an important role in this reaction, and only strongly reducing *fac*-Ir(ppy)$_3$ can efficiently promote this transformation. For electron-rich benzoic acids, the increase in catalyst loading and reaction time were usually required due to their relatively low reactivity. Under optimal conditions, a wide range of 3, 3-disubstituted 2-oxindoles was obtained in a mild manner. The synthetic value was described by the straightforward synthesis of hexahydropyrrolo[2,3-b]indole unit **66e**, which exists widely in many natural products. Similar strategies have been further applied in the hydroacylation of alkenes, reduction of carboxylic acids and deoxygenative radical cyclization [75–78].

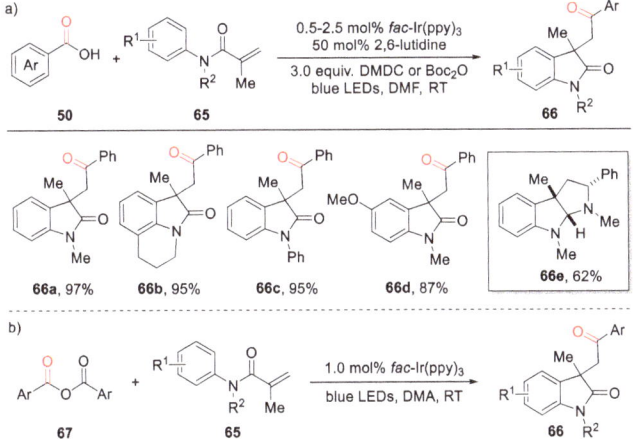

Scheme 18. (a) Acylarylation reaction of aromatic carboxylic acids and methacrylamides; (b) Acyl radical-mediated difunctionalization of olefins.

Rather than the use of in situ formed anhydride intermediates, Wallentin and co-workers directly employed symmetric anhydrides as acyl radical precursors for the difunctionalization of olefins (Scheme 18b) [79]. Two different types of reactions have been developed, including radical acylarylation of N-arylacrylamides and acylation/semipinacol rearrangement of allylic alcohol derivatives. This protocol has been established as a powerful entry to the construction of oxindoles and 1,4-diketones under mild conditions.

A plausible acyl radical addition/intramolecular cyclization mechanism is proposed. In the presence of 2,6-lutidine, the in situ generation of an anhydride intermediate was achieved from benzoic acid and dimethyl dicarbonate (DMDC), which can be reduced by the photoexcited Ir^{III}*-catalyst ($E_{1/2}$ (Ir^{IV}/*Ir^{III}) = −1.73V vs SCE) to form the key acyl radical **50-C** (Scheme 19). Subsequently, the radical **50-C** selectively reacts with olefin **65** to give a C-radical intermediate **65-A**. Finally, the radical **65-A** undergoes a single-electron oxidation and aromatization sequence, producing the oxindole products **66** along with the ground-state of the photocatalyst.

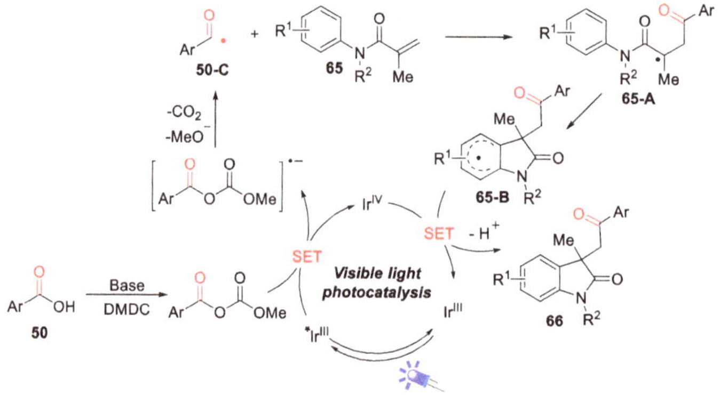

Scheme 19. Plausible mechanism for the acyl radical-mediated acylarylation reaction.

In addition, acyl thioesters are readily available and could be served as a source of acyl radicals. The group of Gryko discovered an unprecedented vitamin B12-catalysed Giese-type acylation of electron-deficient olefins by using 2-S-pyridyl thioesters as acyl radical precursors [80]. An indirect approach to generate acyl radicals from thioesters has been developed by McErlean and co-workers (Scheme 20) [81]. This reaction avoids the use of organo-tin reagent and high-energy UV light irradiation. Under mild photocatalytic conditions, diverse chromanones and indanone derivatives can be synthesized in moderate yields. The synthesis of clinical agent donepezil **72** further demonstrates the potential utility of this reaction.

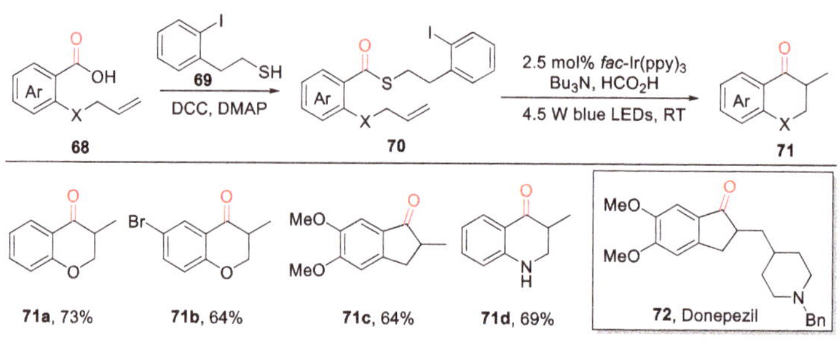

Scheme 20. Acyl thioesters-mediated acyl radical intramolecular cylization reaction.

Acyl chlorides are abundant and highly active acyl radical precursors. In 2017, Xu et al. developed a radical cascade reaction of N-methyl-N-phenylmethacrylamides with aroyl chlorides for the synthesis of various quaternary 3,3-dialkyl 2-oxindole derivatives (Scheme 21a) [82]. In this reaction, aliphatic acyl chlorides have been unsuccessful substrates due to their low reduction potentials. The same group further expanded this concept to the cascade reaction of 1,7-enynes with aroyl chlorides, providing various fused pyran derivatives (Scheme 21b) [83]. This reaction could be scaled up to 4 mmol, albeit with a slightly lower yield.

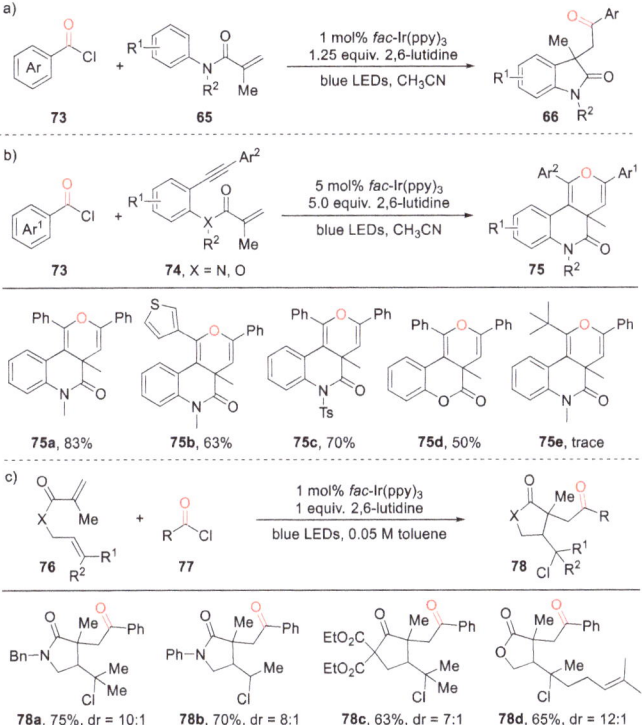

Scheme 21. (a) Visible light photocatalytic cascade reaction of N-methyl-N-phenylmethacrylamides with aroyl chlorides; (b) Cascade reaction of 1,7-enynes with aroyl chlorides; (c) Aroylchlorination reaction of 1,6-dienes.

Recently, Xu and co-workers established a novel photocatalytic aroylchlorination reaction of 1,6-dienes for the synthesis of highly valuable polysubstituted pyrrolidines (Scheme 21c) [84]. Two C–C bonds and one C–Cl bond can be rapidly constructed in a one-pot process. The rational choice of photocatalysts was found to be the key factor of this reaction. Only fac-Ir(ppy)$_3$ proved to be capable of promoting this reaction. Other commonly used photocatalysts such as Ru(bpy)$_3$Cl$_2$, eosin Y and Ir(ppy)$_2$(dtbbpy)PF$_6$ were ineffective. Consistent with their previous works, alkyl chlorides were not compatible with this condition. In addition, Oh et al. reported a facile Friedel–Crafts acylation of alkenes, allowing synthesis of various β-chloroketones under mild photocatalytic conditions [85].

A broad array of 3-acylspiro[4,5]-trienone scaffolds was synthesized by Tang's group through a photocatalytic tandem reaction of N-(p-methoxyaryl)propiolamides with acyl chlorides (Scheme 22a) [86]. This reaction features a broad scope with high selectivity. It was found that a high temperature of 100 °C and 2.0 equivalent of H$_2$O are essential for this reaction. ^{18}O-labeled experiments indicated that the oxygen atom of the newly generated carbonyl group mainly comes from H$_2$O. The same strategy has been further applied for the preparation of 3-acylcoumarins via acyl radical-mediated cyclization of alkynoates by visible light photoredox catalysis (Scheme 22b) [87].

Very recently, Tang et al. developed an acyl radical cyclization of N-propargylindoles for the construction of various important pyrrolo[1,2-*a*]indole skeletons [88]. An acyl radical proved to be the key intermediate in this reaction.

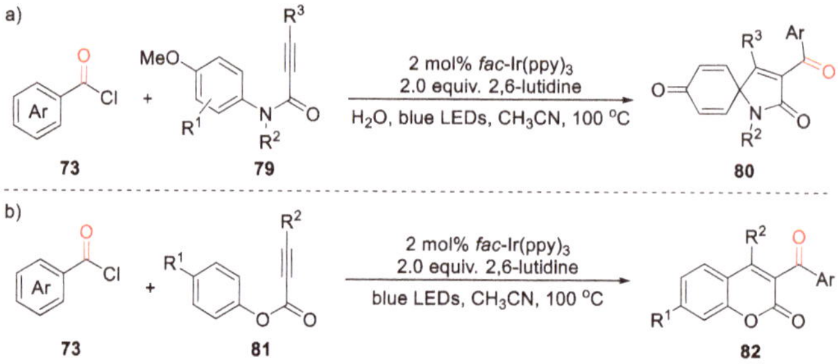

Scheme 22. (a) Acyl radical cascade cylization of N-(*p*-methoxyaryl)propiolamides; (b) Cylization of alkynoates.

Recently, Xuan and Wang et al. developed an elegant photocatalytic acyl radical cyclization of 2-(allyloxy)-benzaldehydes with aroyl chlorides (Scheme 23a) [89]. The unactivated C=C bonds acted as acyl radical acceptors in this process. It was observed that the base is crucial to the reaction efficiency. 2,6-lutidine turned out to be the best base. Under the standard conditions, a series of 2-(allyloxy)-benzaldehydes reacted smoothly with aroyl chlorides to give various chroman-4-one skeletons in moderate yields. More importantly, the chromanone products could be easily converted to other important heterocycles in one-step process, such as benzofuranone and 2-phenyl-4*H*-thieno[3,2-c]chromene.

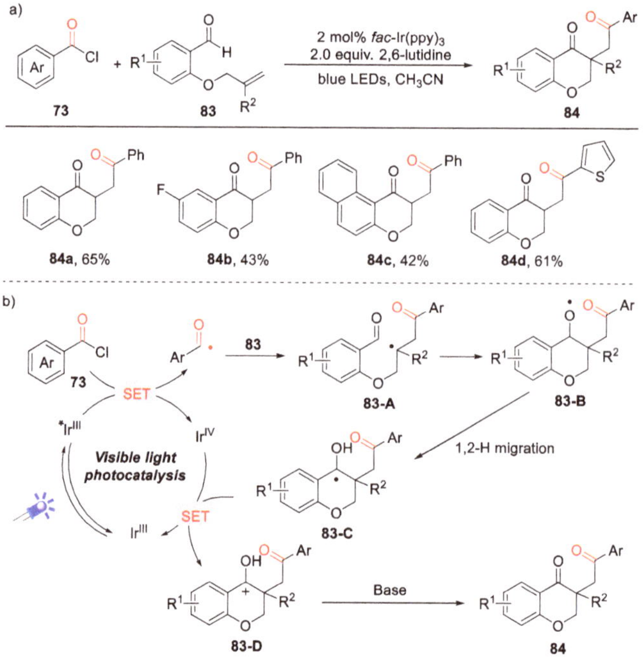

Scheme 23. (a) Radical cyclization of 2-(allyloxy)-benzaldehydes with aroyl chlorides; (b) Proposed mechanism.

A series of fluorescence quenching experiments were investigated to gain insights into the reaction mechanism, which indicated that the excited state of photocatalyst was oxidatively quenched by benzoyl chloride. As described in Scheme 23b, under blue-light irradiation, benzoyl chloride is initially reduced by the photoexcited Ir^{III}* via a single electron transfer, producing the key acyl radical intermediate (Scheme 23b). Then, the selective addition of acyl radical to the unactivated C=C bond affords a C-based radical **83-A**, which then reacts with the carbonyl group to give the O-centered radical intermediate **83-B**. Radical **83-B** undergoes a 1,2-H migration/SET oxidation sequence to give the carbon cation intermediate **83-D**, finishing the visible light photocatalytic cycle. Finally, under basic conditions, intermediate **83-D** proceeds a deprotonation process to give the chroman-4-one product.

5. Miscellaneous Radical Carbonylation

The single-electron oxidative decarboxylation of α-keto acids provides an alternative method to form acyl radicals [90–92]. In 2014, Lei and Lan et al. reported an interesting decarboxylative amidation of α-keto acids (Scheme 24a) [93]. One impressive feature of this transformation is that O_2 worked as the oxidant. A variety of α-keto acids reacted well with anilines, furnishing the final products in generally good yields. For aliphatic amines, 5–10 equivalents of amines were required to achieve good yields. More significantly, this reaction can be further utilized in the construction of N-containing heterocycles such as benzothiazole, benzoxazole and benzimidazole. Visible light irradiation of Ru(II)-catalyst delivers the excited Ru(II)* species via metal-to-ligand charge transfer (MLCT), which is reductively quenched by an amine **86** to give intermediate **86-A**. Then, O_2 acts as an oxidant to regenerate the ground state of Ru(II)-catalyst along with the formation of the superoxide radical anion. The superoxide radical anion further reacts with **85-A** to give radical intermediate **85-B**, which undergoes a decarboxylative process to afford acyl radical **86-B**. Subsequently, this acyl radical reacts with an amine to generate intermediate **86-C**, which then undergoes a SET process to give the amide product. Chu et al. disclosed an interesting palladium-catalyzed decarboxylation coupling/intramolecular cyclization sequence for the formation of 4-aryl-2-quinolinone derivatives (Scheme 24b) [94]. The author further elaborated its potential applicability in the rapid synthesis of the hepatitis B virus (HBV) inhibitor in an atom economy manner.

In 2016, Wang et al. demonstrated an elegant hypervalent iodine (BI–OAc)-mediated decarboxylative cyclization reaction under photocatalyst- and oxidant-free conditions (Scheme 25a) [95]. The reaction produces various oxindoles in moderate to good yields. Mechanistic studies suggested that a cascade decarbonylation, radical addition and cyclization pathway were involved in this reaction (Scheme 25b). The reaction of α-keto acid with BI–OAc results in the hypervalent iodine intermediate **91-A**. Under blue LED irradiation, the homolytic cleavage of **91-A** generates iodanyl radical **91-B** and acyl radical **50-C**. Then, the acyl radical further reacts with acrylamide **65** to give intermediate **65-B**, followed by a hydrogen atom abstraction process to afford the desired product **66** along with the formation of intermediate **91-C**. The intermediate **91-C** may react with α-keto acid to give the hypervalent iodine intermediate **91-A** for the next catalytic cycle.

Using $(NH_4)_2S_2O_8$ as an oxidant, Hu, Huo and Su et al. developed an eosin B-catalyzed decarboxylative cyclization of N-methacryloylbenzamides, providing a wide range of acylated isoquinolines derivatives [96]. A similar concept was applied for the formation of 2-acylindoles via a practical decarboxylative cyclization of 2-alkenylarylisocyanides with α-keto acids (Scheme 26a) [97]. Very recently, Prabhu et al. developed a decarboxylative acylation of electron-deficient heteroarenes in the presence of $Na_2S_2O_8$ (Scheme 26b) [98].

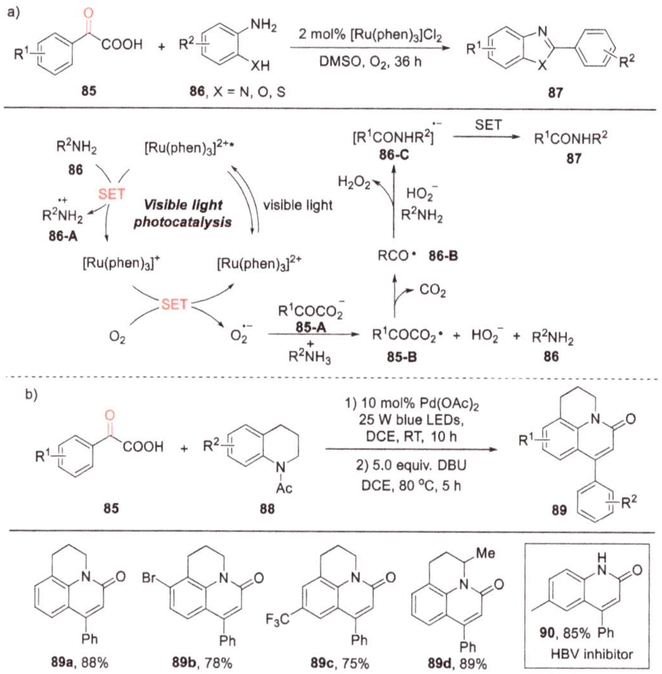

Scheme 24. (**a**) Radical decarboxylative functionalization of α-keto acids; (**b**) Decarboxylation coupling/ intramolecular cyclization sequence.

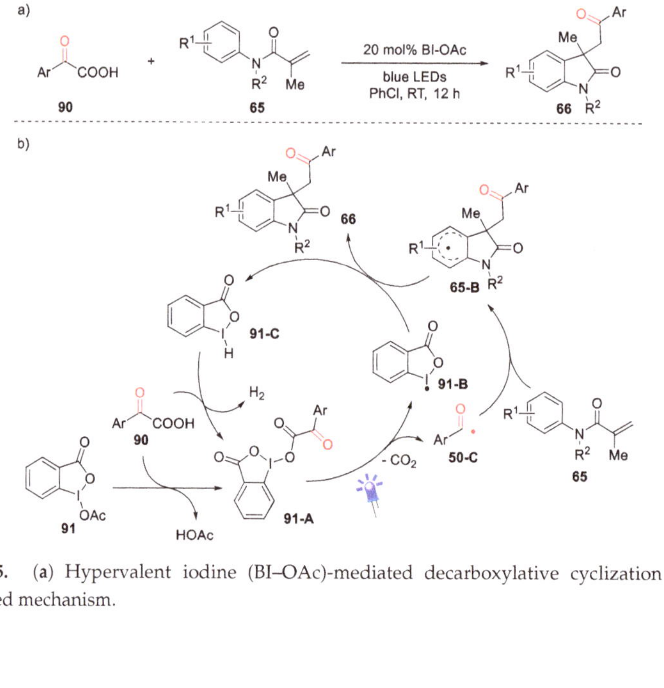

Scheme 25. (**a**) Hypervalent iodine (BI–OAc)-mediated decarboxylative cyclization reaction; (**b**) Proposed mechanism.

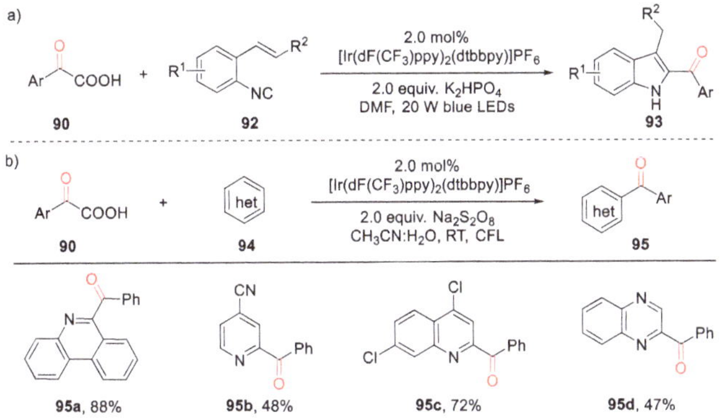

Scheme 26. (a) Decarboxylative cyclization of 2-alkenylarylisocyanides with α-keto acids; (b) Decarboxylative acylation of electron-deficient heteroarenes.

In addition, carbamoyl radicals are important radical species that have been widely used in cascade cyclization reactions for the construction of N-containing heterocycles [99]. In 2018, Feng et al. described an oxidative decarboxylation of oxamic acids to generate carbamoyl radical species (Scheme 27a) [100]. The carbamoyl radicals then react with electron-deficient alkenes for the synthesis of a range of 3,4-dihydroquinolin-2(1H)-ones.

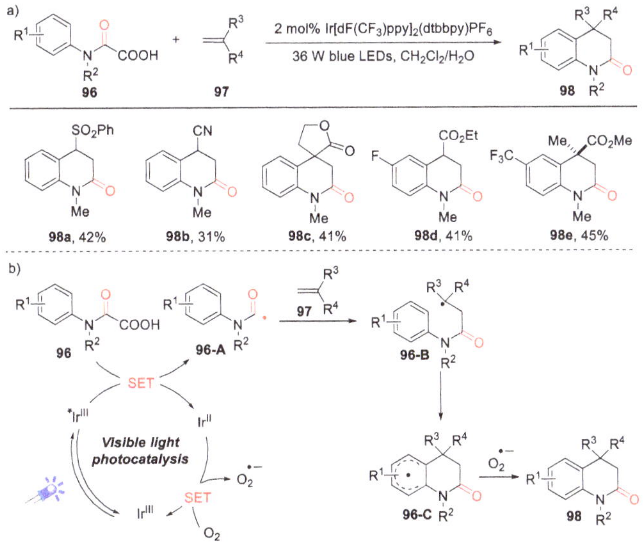

Scheme 27. (a) Photocatalytic decarboxylative functionalization of oxamic acids; (b) Proposed mechanism.

A plausible catalytic cycle is proposed in Scheme 27b. The reaction starts with a single-electron oxidation of oxamic acid by photoexcited catalyst Ir^{III}*, delivering the key carbamoyl radical **96-A** and Ir^{II} species. Intermediate **96-A** reacts with an electron-deficient alkene **97** to give C-based radical **96-B**, which undergoes a rapid intramolecular cyclization to form radical **96-C**. At the same time, the Ir^{II} catalyst can be oxidized by O$_2$ to regenerate the ground state of photocatalyst, along with the formation of oxygen radical anion. Finally, the H-atom abstraction from intermediate **96-C** by oxygen radical anion delivers 3,4-dihydroquinolin-2(1H)-one products.

In addition, Donald, Taylor and co-workers discovered a reductive decarboxylation process for the construction of 3,4-dihydroquinolin-2(1H)-ones (Scheme 28) [101]. In this reaction, the bench-stable and readily prepared N-hydroxyphthalimido oxamides were used as the carbamoyl radical precursors. A wide range of electron-deficient alkenes were successful radical acceptors, such as methyl methacrylate, ethyl vinyl ketone and acrylonitrile. Gratifyingly, the current reaction can be further applied for the assembly of some spirocyclic lactone lactams, which provides an important entry to biologically important spirocycles.

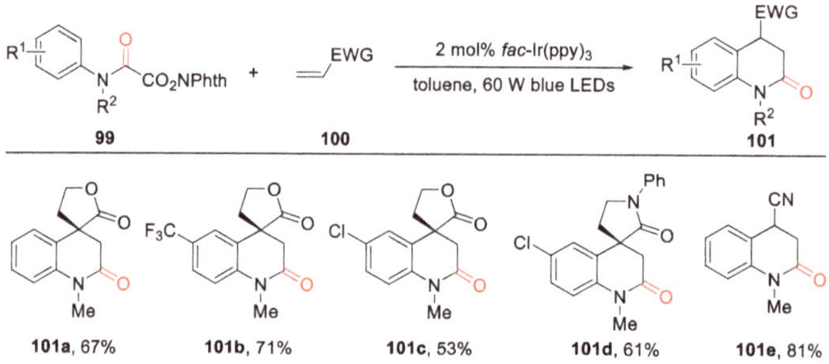

Scheme 28. Reductive decarboxylation of N-hydroxyphthalimido oxamides.

Recently, Wu et al. demonstrated a novel C–C bond activation strategy for the generation of acyl radicals for the first time (Scheme 29) [102]. Under mild photocatalytic conditions, the β-C–C bond fragment of oxime esters leads to a range of aryl acyl radicals in high efficiency. More strikingly, the relatively unstable aliphatic acyl radicals can be successfully formed under the same catalytic system. The newly generated acyl radicals can react with diverse Michael acceptors, such as acrylamides, amines and isonitrile to furnish the desired heterocycles and other linear carbonyl compounds in good yields. A proposed mechanism is depicted in Scheme 30. Under photocatalytic conditions, oxime ester undergoes a fast SET reduction and β-C–C bond homolysis to deliver acyl radical species **102-A** with the elimination of a CH_3CN molecule. The selective addition of radical **102-A** to acrylamide results in a C-radical **103-A**, followed by an intramolecular cyclization/single-electron oxidative aromatization cascade to afford the final products and complete the photocatalytic cycle.

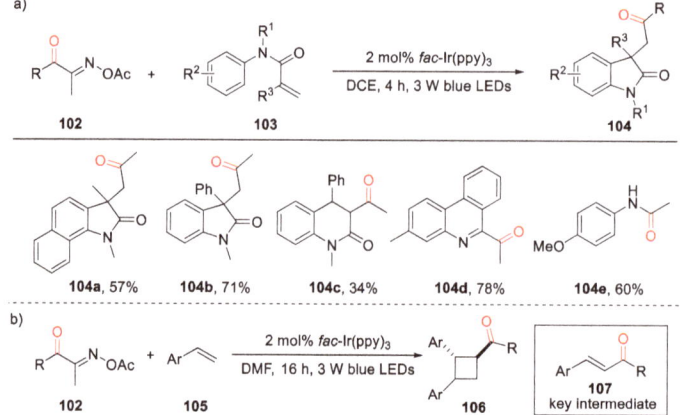

Scheme 29. (a) Photocatalytic C–C bond activation of oxime esters; (b) Photocatalytic [2+2] dimerization reaction.

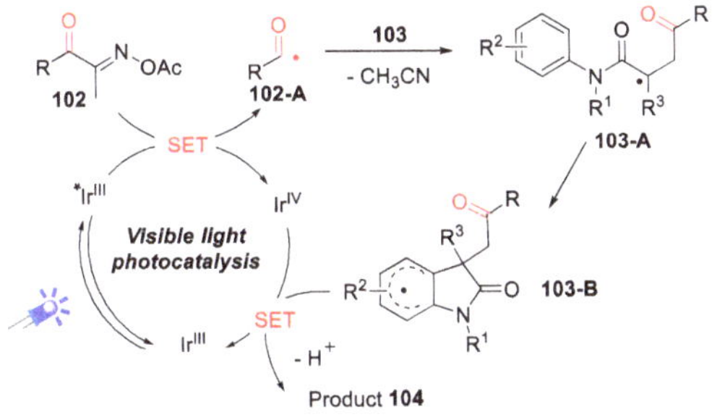

Scheme 30. Proposed mechanism of C–C bond activation in oxime esters.

Surprisingly, the unexpected cyclobutanes could be obtained as the major products via a sequential SET activation and energy transfer (ET) process by using styrenes as acyl radical acceptors in DMF (Scheme 29b). The in situ generated aryl enone 107 proved to be the key intermediate, which undergoes a photocatalytic [2+2] dimerization to give the *anti*-cyclobutane products.

6. Conclusions

Acyl radical-meditated carbonylation has been esteemed as a powerful tool for the efficient construction of a wide range of high-value oxygen and nitrogen heterocycles. Under mild visible light photocatalytic conditions, these reactive acyl radical species can be conveniently generated from diverse acyl radical precursors. Outstanding features of these carbonylative reactions include mild conditions, good functional group tolerance, broad scope and high degree of regioselectivity. Significantly, these reactions have a promising potential in the concise synthesis of biologically active heterocycles and late-stage modification of natural products.

While significant progress emerged in this field, some important challenges remain and need to be addressed in near future. The reaction of unstable aliphatic acyl radicals is largely unexplored due to the unavoidable decarbonylation process. In addition, the radical acceptors of acyl radicals are mainly limited to electron-deficient species. We believe that the development of dual catalytic systems and design of new substrate types may provide solutions to these problems.

Author Contributions: X.-Q.H.: writing—original draft preparation; Z.-K.L.: writing—review and editing; W.-J.X.: supervision. All authors have read and agree to the published version of the manuscript.

Funding: This research was funded by the National Natural Science Foundation of China (21901258 and 21772053).

Conflicts of Interest: The authors declare no conflict of interest.

References

1. Humphrey, J.M.; Chamberlin, A.R. Chemical Synthesis of Natural Product Peptides: Coupling Methods for the Incorporation of Noncoded Amino Acids into Peptides. *Chem. Rev.* **1997**, *97*, 2243–2266. [CrossRef] [PubMed]
2. Zhu, L.; Ding, G.; Xie, L.; Cao, X.; Liu, J.; Lei, X.; Ma, J. Conjugated Carbonyl Compounds as High-Performance Cathode Materials for Rechargeable Batteries. *Chem. Mater.* **2019**, *31*, 8582–8612. [CrossRef]
3. Mao, B.; Fañanás-Mastral, M.; Feringa, B.L. Catalytic Asymmetric Synthesis of Butenolides and Butyrolactones. *Chem. Rev.* **2017**, *117*, 10502–10566. [CrossRef] [PubMed]
4. Smith, A.M.R.; Hii, K.K. Transition Metal Catalyzed Enantioselective A-Heterofunctionalization of Carbonyl Compounds. *Chem. Rev.* **2011**, *111*, 1637–1656. [CrossRef] [PubMed]
5. Jang, J.; Kim, D.Y. Transition Metal-free Phosphorylation of Vinyl Azides: A Convenient Synthesis of B-Ketophosphine Oxides. *Bull. Korean Chem. Soc.* **2020**, *41*, 370–373. [CrossRef]
6. Zhu, C.; Liu, J.; Li, M.B.; Bäckvall, J.E. Palladium-catalyzed Oxidative Dehydrogenative Carbonylation Reactions Using Carbon Monoxide and Mechanistic Overviews. *Chem. Soc. Rev.* **2020**, *49*, 341–353. [CrossRef]

7. Zhang, S.; Neumann, H.; Beller, M. Synthesis of α,β-unsaturated Carbonyl Compounds by Carbonylation Reactions. *Chem. Soc. Rev.* **2020**, *49*, 3187–3210. [CrossRef]
8. Ryu, I. Radical Carboxylations of Iodoalkanes and Saturated Alcohols Using Carbon Monoxide. *Chem. Soc. Rev.* **2001**, *30*, 16–25. [CrossRef]
9. Ryu, I. New Approaches in Radical Carbonylation Chemistry: Fluorous Applications and Designed Tandem Processes by Species-Hybridization with Anions and Transition Metal Species. *Chem. Record* **2002**, *2*, 249–258. [CrossRef]
10. Sumino, S.; Fusano, A.; Fukuyama, T.; Ryu, I. Carbonylation Reactions of Alkyl Iodides through the Interplay of Carbon Radicals and Pd Catalysts. *Acc. Chem. Res.* **2014**, *47*, 1563–1574. [CrossRef]
11. Peng, J.B.; Wu, F.P.; Wu, X.F. First-Row Transition-Metal-Catalyzed Carbonylative Transformations of Carbon Electrophiles. *Chem. Rev.* **2018**, *119*, 2090–2127. [CrossRef] [PubMed]
12. Wu, X.F.; Fang, X.; Wu, L.; Jackstell, R.; Neumann, H.; Beller, M. Transition-Metal-Catalyzed Carbonylation Reactions of Olefins and Alkynes: A Personal Account. *Acc. Chem. Res.* **2014**, *47*, 1041–1053. [CrossRef] [PubMed]
13. Kunin, A.J.; Eisenberg, R. Photochemical Carbonylation of Benzene by Iridium(I) and Rhodium(I) Square-Planar Complexes. *Organometallics* **1988**, *7*, 2124–2129. [CrossRef]
14. Ferguson, R.R.; Crabtree, R.H. Mercury-Photosensitized Sulfination, Hydrosulfination, and Carbonylation of Hydrocarbons: Alkane and Alkene Conversion to Sulfonic Acids, Ketones, and Aldehydes. *J. Org. Chem.* **1991**, *56*, 5503–5510. [CrossRef]
15. Sakakura, T.; Sodeyama, T.; Sasaki, K.; Wada, K.; Tanaka, M. Carbonylation of Hydrocarbons via C-H Activation Catalyzed by $RhCl(CO)(PMe_3)_2$ under Irradiation. *J. Am. Chem. Soc.* **1990**, *112*, 7221–7229. [CrossRef]
16. Jaynes, B.S.; Hill, C.L. Radical Carbonylation of Alkanes via Polyoxotungstate Photocatalysis. *J. Am. Chem. Soc.* **1995**, *117*, 4704–4705. [CrossRef]
17. Pitzer, L.; Sandfort, F.; Strieth-Kalthoff, F.; Glorius, F. Carbonyl–Olefin Cross-Metathesis Through a Visible-Light-Induced 1,3-Diol Formation and Fragmentation Sequence. *Angew. Chem. Int. Ed.* **2018**, *57*, 16219–16223. [CrossRef]
18. Yu, X.Y.; Zhao, Q.Q.; Chen, J.; Xiao, W.-J.; Chen, J.-R. When Light Meets Nitrogen-Centered Radicals: From Reagents to Catalysts. *Acc. Chem. Res.* **2020**, *53*, 1066–1083. [CrossRef]
19. Festa, A.A.; Voskressensky, L.G.; Van der Eycken, E.V. Visible Light-mediated Chemistry of Indoles and Related Heterocycles. *Chem. Soc. Rev.* **2019**, *48*, 4401–4423. [CrossRef]
20. Silvi, M.; Melchiorre, P. Enhancing the Potential of Enantioselective Organocatalysis with Light. *Nature* **2018**, *554*, 41–49. [CrossRef]
21. Li, W.; Xu, W.; Xie, J.; Yu, S.; Zhu, C. Distal Radical Migration Strategy: An Emerging Synthetic Means. *Chem. Soc. Rev.* **2018**, *47*, 654–667. [CrossRef]
22. Kärkäs, M.D.; Porco, J.A.; Stephenson, C.R.J. Photochemical Approaches to Complex Chemotypes: Applications in Natural Product Synthesis. *Chem. Rev.* **2016**, *116*, 9683–9747. [CrossRef] [PubMed]
23. Skubi, K.L.; Blum, T.R.; Yoon, T.P. Dual Catalysis Strategies in Photochemical Synthesis. *Chem. Rev.* **2016**, *116*, 10035–10074. [CrossRef]
24. Prier, C.K.; Rankic, D.A.; MacMillan, D.W.C. Visible Light Photoredox Catalysis with Transition Metal Complexes: Applications in Organic Synthesis. *Chem. Rev.* **2013**, *113*, 5322–5363. [CrossRef] [PubMed]
25. Fu, M.C.; Shang, R.; Zhao, B.; Wang, B.; Fu, Y. Photocatalytic Decarboxylative Alkylations Mediated by Triphenylphosphine and Sodium Iodide. *Science* **2019**, *363*, 1429–1434. [CrossRef] [PubMed]
26. Strieth-Kalthoff, F.; James, M.J.; Teders, M.; Pitzer, L.; Glorius, F. Energy Transfer Catalysis Mediated by Visible Light: Principles, Applications, Directions. *Chem. Soc. Rev.* **2018**, *47*, 7190–7202. [CrossRef] [PubMed]
27. Yi, H.; Zhang, G.; Wang, H.; Huang, Z.; Wang, J.; Singh, A.K.; Lei, A. Recent Advances in Radical C–H Activation/Radical Cross-Coupling. *Chem. Rev.* **2017**, *117*, 9016–9085. [CrossRef] [PubMed]
28. Qiu, G.; Lai, L.; Cheng, J.; Wu, J. Recent Advances in the Sulfonylation of Alkenes with the Insertion of Sulfur Dioxide via Radical Reactions. *Chem. Commun.* **2018**, *54*, 10405–10414. [CrossRef]
29. Chen, J.-R.; Hu, X.-Q.; Lu, L.-Q.; Xiao, W.-J. Visible Light Photoredox-controlled Reactions of N-radicals and Radical Ions. *Chem. Soc. Rev.* **2016**, *45*, 2044–2056. [CrossRef]

30. Chen, J.-R.; Hu, X.-Q.; Lu, L.-Q.; Xiao, W.-J. Exploration of Visible-Light Photocatalysis in Heterocycle Synthesis and Functionalization: Reaction Design and Beyond. *Acc. Chem. Res.* **2016**, *49*, 1911–1923. [CrossRef]
31. Banerjee, A.; Lei, Z.; Ngai, M.Y. Acyl Radical Chemistry via Visible-Light Photoredox Catalysis. *Synthesis* **2018**, *51*, 303–333. [PubMed]
32. Zhao, S.; Mankad, N.P. Metal-catalysed Radical Carbonylation Reactions. *Catal. Sci. Technol.* **2019**, *9*, 3603–3613. [CrossRef]
33. Matsubara, H.; Kawamoto, T.; Fukuyama, T.; Ryu, I. Applications of Radical Carbonylation and Amine Addition Chemistry: 1,4-Hydrogen Transfer of 1-Hydroxylallyl Radicals. *Acc. Chem. Res.* **2018**, *51*, 2023–2035. [CrossRef]
34. Peng, J.B.; Geng, H.Q.; Wu, X.F. The Chemistry of CO: Carbonylation. *Chem* **2019**, *5*, 526–552. [CrossRef]
35. Zhang, H.; Shi, R.; Ding, A.; Lu, L.; Chen, B.; Lei, A. Transition-Metal-Free Alkoxycarbonylation of Aryl Halides. *Angew. Chem. Int. Ed.* **2012**, *51*, 12542–12545. [CrossRef]
36. Guo, W.; Lu, L.-Q.; Wang, Y.; Wang, Y.N.; Chen, J.-R.; Xiao, W.-J. Metal-Free, Room-Temperature, Radical Alkoxycarbonylation of Aryldiazonium Salts Through Visible-Light Photoredox Catalysis. *Angew. Chem. Int. Ed.* **2014**, *54*, 2265–2269. [CrossRef] [PubMed]
37. Majek, M.; Jacobi von Wangelin, A. Metal-Free Carbonylations by Photoredox Catalysis. *Angew. Chem. Int. Ed.* **2014**, *54*, 2270–2274. [CrossRef]
38. Gu, L.; Jin, C.; Liu, J. Metal-free, Visible-light-mediated Transformation of Aryl Diazonium Salts and (hetero)arenes: An Efficient Route to Aryl Ketones. *Green Chem.* **2015**, *17*, 3733–3736. [CrossRef]
39. Zhang, H.T.; Gu, L.J.; Huang, X.Z.; Wang, R.; Jin, C.; Li, G.P. Synthesis of Indol-3-yl Aryl Ketones Through Visible-light-mediated Carbonylation. *Chin. Chem. Lett.* **2016**, *27*, 256–260. [CrossRef]
40. Li, X.; Liang, D.; Huang, W.; Zhou, H.; Li, Z.; Wang, B.; Ma, Y.; Wang, H. Visible Light-induced Carbonylation of Indoles with Arylsulfonyl Chlorides and CO. *Tetrahedron* **2016**, *72*, 8442–8448. [CrossRef]
41. Micic, N.; Polyzos, A. Radical Carbonylation Mediated by Continuous-Flow Visible-Light Photocatalysis: Access to 2,3-Dihydrobenzofurans. *Org. Lett.* **2018**, *20*, 4663–4666. [CrossRef]
42. Zhou, Q.-Q.; Guo, W.; Ding, W.; Wu, X.; Chen, X.; Lu, L.-Q.; Xiao, W.-J. Decarboxylative Alkynylation and Carbonylative Alkynylation of Carboxylic Acids Enabled by Visible-Light Photoredox Catalysis. *Angew. Chem. Int. Ed.* **2015**, *54*, 11196–11199. [CrossRef] [PubMed]
43. Jiang, X.; Zhang, M.; Xiong, W.; Lu, L.-Q.; Xiao, W.-J. Deaminative (Carbonylative) Alkyl-Heck-type Reactions Enabled by Photocatalytic C–N Bond Activation. *Angew. Chem. Int. Ed.* **2019**, *58*, 2402–2406. [CrossRef] [PubMed]
44. Basch, C.H.; Liao, J.; Xu, J.; Piane, J.J.; Watson, M.P. Harnessing Alkyl Amines as Electrophiles for Nickel-Catalyzed Cross Couplings via C–N Bond Activation. *J. Am. Chem. Soc.* **2017**, *139*, 5313–5316. [CrossRef] [PubMed]
45. Yue, H.; Zhu, C.; Shen, L.; Geng, Q.; Hock, K.J.; Yuan, T.; Cavallo, L.; Rueping, M. Nickel-catalyzed C–N Bond Activation: Activated Primary Amines as Alkylating Reagents in Reductive Cross-coupling. *Chem. Sci.* **2019**, *10*, 4430–4435. [CrossRef]
46. Wu, J.; He, L.; Noble, A.; Aggarwal, V.K. Photoinduced Deaminative Borylation of Alkylamines. *J. Am. Chem. Soc.* **2018**, *140*, 10700–10704. [CrossRef] [PubMed]
47. Lu, B.; Cheng, Y.; Chen, L.Y.; Chen, J.-R.; Xiao, W.-J. Photoinduced Copper-Catalyzed Radical Aminocarbonylation of Cycloketone Oxime Esters. *ACS Catal.* **2019**, *9*, 8159–8164. [CrossRef]
48. Torres, G.M.; Liu, Y.; Arndtsen, B.A. A Dual Light-driven Palladium Catalyst: Breaking the Barriers in Carbonylation Reactions. *Science* **2020**, *368*, 318–323. [CrossRef]
49. Cartier, A.; Levernier, E.; Dhimane, A.; Fukuyama, T.; Ollivier, C.; Ryu, I.; Fensterbank, L. Synthesis of Aliphatic Amides Through a Photoredox Catalyzed Radical Carbonylation Involving Organosilicates as Alkyl Radical Precursors. *Adv. Synth. Catal.* **2020**, *362*, 2254–2259. [CrossRef]
50. Liu, K.; Zou, M.; Lei, A. Aerobic Oxidative Carbonylation of Enamides by Merging Palladium with Photoredox Catalysis. *J. Org. Chem.* **2016**, *81*, 7088–7092. [CrossRef]
51. Chatgilialoglu, C.; Crich, D.; Komatsu, M.; Ryu, I. Chemistry of Acyl Radicals. *Chem. Rev.* **1999**, *99*, 1991–2070. [CrossRef] [PubMed]
52. Cheng, P.; Qing, Z.; Liu, S.; Liu, W.; Xie, H.; Zeng, J. Regiospecific Minisci Acylation of Phenanthridine via Thermolysis or Photolysis. *Tetrahedron Lett.* **2014**, *55*, 6647–6651. [CrossRef]

53. Capaldo, L.; Ravelli, D. Hydrogen Atom Transfer (HAT): A Versatile Strategy for Substrate Activation in Photocatalyzed Organic Synthesis. *Eur. J. Org. Chem.* **2017**, *2017*, 2056–2071. [CrossRef] [PubMed]
54. Li, J.; Wang, D.Z. Visible-Light-Promoted Photoredox Syntheses of α, β-Epoxy Ketones from Styrenes and Benzaldehydes Under Alkaline Conditions. *Org. Lett.* **2015**, *17*, 5260–5263. [CrossRef]
55. de Souza, G.F.P.; Bonacin, J.A.; Salles, A.G., Jr. Visible-Light-Driven Epoxyacylation and Hydroacylation of Olefins Using Methylene Blue/Persulfate System in Water. *J. Org. Chem.* **2018**, *83*, 8331–8340. [CrossRef]
56. Voutyritsa, E.; Kokotos, C.G. Green Metal-Free Photochemical Hydroacylation of Unactivated Olefins. *Angew. Chem. Int. Ed.* **2020**, *59*, 1735–1741. [CrossRef]
57. Zhao, X.; Li, B.; Xia, W. Visible-Light-Promoted Photocatalyst-Free Hydroacylation and Diacylation of Alkenes Tuned by $NiCl_2 \cdot DME$. *Org. Lett.* **2020**, *22*, 1056–1061. [CrossRef]
58. Goti, G.; Bieszczad, B.; Vega-Peñaloza, A.; Melchiorre, P. Stereocontrolled Synthesis of 1,4-Dicarbonyl Compounds by Photochemical Organocatalytic Acyl Radical Addition to Enals. *Angew. Chem. Int. Ed.* **2019**, *58*, 1213–1217. [CrossRef]
59. Jung, S.; Kim, J.; Hong, S. Visible Light-Promoted Synthesis of Spiroepoxy Chromanone Derivatives via a Tandem Oxidation/Radical Cyclization/Epoxidation Process. *Adv. Synth. Catal.* **2017**, *359*, 3945–3949. [CrossRef]
60. Kawaai, K.; Yamaguchi, T.; Yamaguchi, E.; Endo, S.; Tada, N.; Ikari, A.; Itoh, A. Photoinduced Generation of Acyl Radicals from Simple Aldehydes, Access to 3-Acyl-4-arylcoumarin Derivatives, and Evaluation of Their Antiandrogenic Activities. *J. Org. Chem.* **2018**, *83*, 1988–1996. [CrossRef]
61. Lux, M.; Klussmann, M. Additions of Aldehyde-Derived Radicals and Nucleophilic N-Alkylindoles to Styrenes by Photoredox Catalysis. *Org. Lett.* **2020**, *22*, 3697–3701. [CrossRef]
62. Zhang, Y.; Ji, P.; Dong, Y.; Wei, Y.; Wang, W. Deuteration of Formyl Groups via a Catalytic Radical H/D Exchange Approach. *ACS Catal.* **2020**, *10*, 2226–2230. [CrossRef]
63. Ackermann, L. Carboxylate-Assisted Transition-Metal-Catalyzed C–H Bond Functionalizations: Mechanism and Scope. *Chem. Rev.* **2011**, *111*, 1315–1345. [CrossRef] [PubMed]
64. Gooßen, L.; Rodríguez, N.; Gooßen, K. Carboxylic Acids as Substrates in Homogeneous Catalysis. *Angew. Chem. Int. Ed.* **2008**, *47*, 3100–3120. [CrossRef]
65. Font, M.; Quibell, J.M.; Perry, G.J.P.; Larrosa, I. The Use of Carboxylic Acids as Traceless Directing Groups for Regioselective C–H Bond Functionalisation. *Chem. Commun.* **2017**, *53*, 5584–5597. [CrossRef] [PubMed]
66. Hu, X.-Q.; Liu, Z.-K.; Hou, Y.-X.; Gao, Y. Single Electron Activation of Aryl Carboxylic Acids. *iScience* **2020**, *23*, 101266. [CrossRef]
67. Zhang, M.; Xie, J.; Zhu, C. A General Deoxygenation Approach for Synthesis of Ketones from Aromatic Carboxylic Acids and Alkenes. *Nat. Commun.* **2018**, *9*, 3517. [CrossRef]
68. Stache, E.E.; Ertel, A.B.; Rovis, T.; Doyle, A.G. Generation of Phosphoranyl Radicals via Photoredox Catalysis Enables Voltage–Independent Activation of Strong C–O Bonds. *ACS Catal.* **2018**, *8*, 11134–11139. [CrossRef]
69. Zhang, M.; Yuan, X.A.; Zhu, C.; Xie, J. Deoxygenative Deuteration of Carboxylic Acids with D_2O. *Angew. Chem. Int. Ed.* **2018**, *58*, 312–316. [CrossRef]
70. Ruzi, R.; Ma, J.; Yuan, X.; Wang, W.; Wang, S.; Zhang, M.; Dai, J.; Xie, J.; Zhu, C. Deoxygenative Arylation of Carboxylic Acids by Aryl Migration. *Chem. Eur. J.* **2019**, *25*, 12724–12729. [CrossRef]
71. Jiang, H.; Mao, G.; Wu, H.; An, Q.; Zuo, M.; Guo, W.; Xu, C.; Sun, Z.; Chu, W. Synthesis of Dibenzocycloketones by Acyl Radical Cyclization from Aromatic Carboxylic Acids Using Methylene Blue as a Photocatalyst. *Green Chem.* **2019**, *21*, 5368–5373. [CrossRef]
72. Martinez Alvarado, J.I.; Ertel, A.B.; Stegner, A.; Stache, E.E.; Doyle, A.G. Direct Use of Carboxylic Acids in the Photocatalytic Hydroacylation of Styrenes To Generate Dialkyl Ketones. *Org. Lett.* **2019**, *21*, 9940–9944. [CrossRef] [PubMed]
73. Guo, Y.Q.; Wang, R.; Song, H.; Liu, Y.; Wang, Q. Visible-Light-Induced Deoxygenation/Defluorination Protocol for Synthesis of γ,γ-Difluoroallylic Ketones. *Org. Lett.* **2020**, *22*, 709–713. [CrossRef] [PubMed]
74. Bergonzini, G.; Cassani, C.; Wallentin, C.J. Acyl Radicals from Aromatic Carboxylic Acids by Means of Visible-Light Photoredox Catalysis. *Angew. Chem. Int. Ed.* **2015**, *54*, 14066–14069. [CrossRef] [PubMed]
75. Pettersson, F.; Bergonzini, G.; Cassani, C.; Wallentin, C.J. Redox-Neutral Dual Functionalization of Electron-Deficient Alkenes. *Chem. Eur. J.* **2017**, *23*, 7444–7447. [CrossRef] [PubMed]
76. Zhang, M.; Ruzi, R.; Xi, J.; Li, N.; Wu, Z.; Li, W.; Yu, S.; Zhu, C. Photoredox-Catalyzed Hydroacylation of Olefins Employing Carboxylic Acids and Hydrosilanes. *Org. Lett.* **2017**, *19*, 3430–3433. [CrossRef] [PubMed]

77. Zhang, M.; Li, N.; Tao, X.; Ruzi, R.; Yu, S.; Zhu, C. Selective Reduction of Carboxylic Acids to Aldehydes with Hydrosilane via Photoredox Catalysis. *Chem. Commun.* **2017**, *53*, 10228–10231. [CrossRef] [PubMed]
78. Ruzi, R.; Zhang, M.; Ablajan, K.; Zhu, C. Photoredox-Catalyzed Deoxygenative Intramolecular Acylation of Biarylcarboxylic Acids: Access to Fluorenones. *J. Org. Chem.* **2017**, *82*, 12834–12839. [CrossRef]
79. Bergonzini, G.; Cassani, C.; Lorimer-Olsson, H.; Hörberg, J.; Wallentin, C.J. Visible-Light-Mediated Photocatalytic Difunctionalization of Olefins by Radical Acylarylation and Tandem Acylation/Semipinacol Rearrangement. *Chem. Eur. J.* **2016**, *22*, 3292–3295. [CrossRef]
80. Ociepa, M.; Baka, O.; Narodowiec, J.; Gryko, D. Light-Driven Vitamin B12-Catalysed Generation of Acyl Radicals from 2-S-Pyridyl Thioesters. *Adv. Synth. Catal.* **2017**, *359*, 3560–3565. [CrossRef]
81. Norman, A.R.; Yousif, M.N.; McErlean, C.S.P. Photoredox-catalyzed Indirect Acyl Radical Generation from Thioesters. *Org. Chem. Front.* **2018**, *5*, 3267–3298. [CrossRef]
82. Xu, S.M.; Chen, J.Q.; Liu, D.; Bao, Y.; Liang, Y.M.; Xu, P.F. Aroyl Chlorides as Novel Acyl Radical Precursors via Visible-light Photoredox Catalysis. *Org. Chem. Front.* **2017**, *4*, 1331–1335. [CrossRef]
83. Li, C.G.; Xu, G.Q.; Xu, P.F. Synthesis of Fused Pyran Derivatives via Visible-Light-Induced Cascade Cyclization of 1,7-Enynes with Acyl Chlorides. *Org. Lett.* **2017**, *19*, 512–515. [CrossRef] [PubMed]
84. Zhao, Q.S.; Xu, G.Q.; Liang, H.; Wang, Z.Y.; Xu, P.F. Aroylchlorination of 1,6-Dienes via a Photoredox Catalytic Atom-Transfer Radical Cyclization Process. *Org. Lett.* **2019**, *21*, 8615–8619. [CrossRef] [PubMed]
85. Patil, D.V.; Kim, H.Y.; Oh, K. Visible Light-Promoted Friedel–Crafts-Type Chloroacylation of Alkenes to B-Chloroketones. *Org. Lett.* **2020**, *22*, 3018–3022. [CrossRef]
86. Liu, Y.; Wang, Q.L.; Zhou, C.S.; Xiong, B.Q.; Zhang, P.L.; Yang, C.A.; Tang, K.W. Visible-Light-Mediated Ipso-Carboacylation of Alkynes: Synthesis of 3-Acylspiro[4,5]trienones from N-(p-Methoxyaryl)propiolamides and Acyl Chlorides. *J. Org. Chem.* **2018**, *83*, 2210–2218. [CrossRef] [PubMed]
87. Liu, Y.; Wang, Q.L.; Zhou, C.S.; Xiong, B.Q.; Zhang, P.L.; Kang, S.J.; Yang, C.A.; Tang, K.W. Visible-light-mediated Cascade Difunctionalization/cyclization of Alkynoates with Acyl Chlorides for Synthesis of 3-acylcoumarins. *Tetrahedron Lett.* **2018**, *59*, 2038–2041. [CrossRef]
88. Liu, Y.; Chen, Z.; Wang, Q.L.; Chen, P.; Xie, J.; Xiong, B.Q.; Zhang, P.L.; Tang, K.W. Visible Light-Catalyzed Cascade Radical Cyclization of N-Propargylindoles with Acyl Chlorides for the Synthesis of 2-Acyl-9H-pyrrolo[1,2-a]indoles. *J. Org. Chem.* **2020**, *85*, 2385–2394. [CrossRef]
89. He, X.; Cai, B.; Yang, Q.; Wang, L.; Xuan, J. Visible-Light-Promoted Cascade Radical Cyclization: Synthesis of 1,4-Diketones Containing Chroman-4-One Skeletons. *Chem. Asian J.* **2019**, *14*, 3269–3273. [CrossRef]
90. Wang, G.Z.; Shang, R.; Cheng, W.M.; Fu, Y. Decarboxylative 1,4-Addition of A-Oxocarboxylic Acids with Michael Acceptors Enabled by Photoredox Catalysis. *Org. Lett.* **2015**, *17*, 4830–4833. [CrossRef]
91. Zhao, J.J.; Zhang, H.H.; Shen, X.; Yu, S. Enantioselective Radical Hydroacylation of Enals with A-Ketoacids Enabled by Photoredox/Amine Cocatalysis. *Org. Lett.* **2019**, *21*, 913–916. [CrossRef] [PubMed]
92. Li, J.; Liu, Z.; Wu, S.; Chen, Y. Acyl Radical Smiles Rearrangement to Construct Hydroxybenzophenones by Photoredox Catalysis. *Org. Lett.* **2019**, *21*, 2077–2080. [CrossRef] [PubMed]
93. Liu, J.; Liu, Q.; Yi, H.; Qin, C.; Bai, R.; Qi, X.; Lan, Y.; Lei, A. Visible-Light-Mediated Decarboxylation/Oxidative Amidation of A-Keto Acids with Amines Under Mild Reaction Conditions Using O_2. *Angew. Chem. Int. Ed.* **2013**, *53*, 502–506. [CrossRef] [PubMed]
94. Wang, C.; Qiao, J.; Liu, X.; Song, H.; Sun, Z.; Chu, W. Visible-Light-Induced Decarboxylation Coupling/Intramolecular Cyclization: A One-Pot Synthesis for 4-Aryl-2-quinolinone Derivatives. *J. Org. Chem.* **2018**, *83*, 1422–1430. [CrossRef]
95. Ji, W.; Tan, H.; Wang, M.; Li, P.; Wang, L. Photocatalyst-free Hypervalent Iodine Reagent Catalyzed Decarboxylative Acylarylation of Acrylamides with A-oxocarboxylic Acids Driven by Visible-light Irradiation. *Chem. Commun.* **2016**, *52*, 1462–1465. [CrossRef]
96. Su, Y.; Zhang, R.; Xue, W.; Liu, X.; Zhao, Y.; Wang, K.H.; Huang, D.; Huo, C.; Hu, Y. Visible-light-promoted Acyl Radical Cascade Reaction for Accessing Acylated Isoquinoline-1,3(2H,4H)-dione Derivatives. *Org. Biomol. Chem.* **2020**, *18*, 1940–1948. [CrossRef]
97. Zhang, X.; Zhu, P.; Zhang, R.; Li, X.; Yao, T. Visible-Light-Induced Decarboxylative Cyclization of 2-Alkenylarylisocyanides with A-Oxocarboxylic Acids: Access to 2-Acylindoles. *J. Org. Chem.* **2020**. [CrossRef]

98. Manna, S.; Prabhu, K.R. Visible-Light-Mediated Direct Decarboxylative Acylation of Electron-Deficient Heteroarenes Using A-Ketoacids. *J. Org. Chem.* **2019**, *84*, 5067–5077. [CrossRef]
99. Petersen, W.F.; Taylor, R.J.K.; Donald, J.R. Photoredox-catalyzed Procedure for Carbamoyl Radical Generation: 3,4-dihydroquinolin-2-one and Quinolin-2-one Synthesis. *Org. Biomol. Chem.* **2017**, *15*, 5831–5845. [CrossRef]
100. Bai, Q.F.; Jin, C.; He, J.Y.; Feng, G. Carbamoyl Radicals via Photoredox Decarboxylation of Oxamic Acids in Aqueous Media: Access to 3,4-Dihydroquinolin-2(1H)-ones. *Org. Lett.* **2018**, *20*, 2172–2175. [CrossRef]
101. Petersen, W.F.; Taylor, R.J.K.; Donald, J.R. Photoredox-Catalyzed Reductive Carbamoyl Radical Generation: A Redox-Neutral Intermolecular Addition–Cyclization Approach to Functionalized 3,4-Dihydroquinolin-2-ones. *Org. Lett.* **2017**, *19*, 874–877. [CrossRef] [PubMed]
102. Fan, X.; Lei, T.; Chen, B.; Tung, C.H.; Wu, L.Z. Photocatalytic C–C Bond Activation of Oxime Ester for Acyl Radical Generation and Application. *Org. Lett.* **2019**, *21*, 4153–4158. [CrossRef] [PubMed]

© 2020 by the authors. Licensee MDPI, Basel, Switzerland. This article is an open access article distributed under the terms and conditions of the Creative Commons Attribution (CC BY) license (http://creativecommons.org/licenses/by/4.0/).

Review

From Alkynes to Heterocycles through Metal-Promoted Silylformylation and Silylcarbocyclization Reactions

Gianluigi Albano [1] and Laura Antonella Aronica [2,*]

1. Dipartimento di Chimica, Università degli Studi di Bari "Aldo Moro", Via Edoardo Orabona 4, 70126 Bari, Italy; gianluigi.albano@uniba.it
2. Dipartimento di Chimica e Chimica Industriale, Università di Pisa, Via Giuseppe Moruzzi 13, 56124 Pisa, Italy
* Correspondence: laura.antonella.aronica@unipi.it

Received: 6 August 2020; Accepted: 29 August 2020; Published: 3 September 2020

Abstract: Oxygen and nitrogen heterocyclic systems are present in a large number of natural and synthetic compounds. In particular, oxa- and aza-silacyclane, tetrahydrofuran, benzofuran, cycloheptadifuranone, cycloheptadipyrrolone, pyrrolidine, lactone, lactam, phthalan, isochromanone, tetrahydroisoquinolinone, benzoindolizidinone, indoline and indolizidine scaffolds are present in many classes of biologically active molecules. Most of these contain a C=O moiety which can be easily introduced using carbonylative reaction conditions. In this field, intramolecular silylformylation and silylcarbocyclization reactions may afford heterocyclic compounds containing a carbonyl functional group together with a vinylsilane moiety which can be further transformed. Considering these two aspects, in this review a detailed analysis of the literature data regarding the application of silylformylation and silylcarbocyclization reactions to the synthesis of several heterocyclic derivatives is reported.

Keywords: silylformylation; silylcarbocyclization; alkynes; *N*-heterocycles; *O*-heterocycles

1. Introduction

The silylformylation reaction of terminal acetylenic compounds [1–5] consists of the simultaneous introduction of a trialkylsilylgroup and a formyl moiety into a carbon–carbon multiple bond (Scheme 1). The reaction takes place with total regio- and stereoselectivity, -CHO and -SiR$_3$ being added *syn* to the triple bond with the formyl group bonded to the carbon atom connected to the alkyl chain.

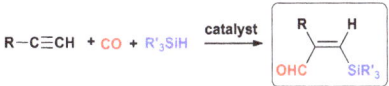

Scheme 1. General scheme of silylformylation reaction.

This reaction represents an extension of the well-known hydroformylation process [6–15], where the H$_2$ molecule is replaced by a hydrosilane. Since the first study of Matsuda et al. that appeared in 1989 [16], the silylformylation of triple bonds has been extensively studied as it provides a direct route to the synthesis of β-silylalkenals. Many different rhodium catalysts have been found to be effective in the alkynes silylformylation. Rh$_4$(CO)$_{12}$ is the most widely used species [16–19], but also Rh(I) [20,21], Rh(II) [22–24] and bimetallic Rh-Co [25–29] complexes were employed. Moreover, Doyle investigated the catalytic activity of Rh$_2$(pfb)$_4$ (perfluorobutyrrate) [30,31] in the silylformylation of terminal alkynes, Alper developed a zwitterionic species, (η^6-C$_6$H$_6$BPh$_3$)$^-$Rh$^+$(1,5-COD) (Rh^{sw}) [32–34] and Aronica et al. [35] showed that rhodium nanocluster, obtained by Metal Vapor Synthesis (MVS) technique could be able to catalyze the silylformylation of linear and branched acetylenes. It is worth noting that silylformylation of alkynes is generally tolerant of many functionalities such as ethers, esters, alcohols, ketones, aldehydes, amines, nitrile, chlorine, bromine and double bonds [18,26,30,33,36,37].

Many applications of the silylformylation of terminal acetylenes have been carried out. In organic synthesis, this process represents an ideal route to many organic compounds because of its high regio and stereoselectivity. It provides β-silylalkenals, which can be easily transformed into silylsubstituted dienes, dienones, α,β-unsaturated ketones and alcohols [35–42], and can be important precursors for the synthesis of more complicated molecules via Peterson olefination [43] or Nazarov-type annulation [44,45]. Finally, fluoride promoted aromatic ring migration from the dimethylarylsilyl moiety to the adjacent carbon atom of the β-silylalkenal yields 2-(arylmethyl)alkanals (Scheme 2) [18,19,46].

Scheme 2. Fluoride-promoted aryl rearrangement of β-silylalkenals: synthesis of 2-(arylmethyl)alkanals.

The "cyclic" version of silylformylation reaction may occur into two different ways: (1) *intramolecular silylformylation* of ω-silylacetylenes, giving the corresponding silacycloalkanes; (2) *silylcarbocyclization* reactions (SiCaC) of suitable alkenynes, involving the formation of cyclic compounds together with the insertion of a silane and a -CHO functional groups. Therefore, the content of this review will be divided into two sections: the first is dedicated to giving a detailed description of intramolecular silylformylation reactions, while the second is centered on the silylcarbocyclization of functionalized acetylenes. In each section we will give particular emphasis to the heterocycles which can be obtained, as well as a special look to the used metal catalysts.

2. Heterocycles Synthesis via Metal-Catalyzed Intramolecular Silylformylation of Alkynes

2.1. The Intramolecular Silylformylation of ω-Silylalkynes: Synthesis of Silacyclanes

The first example of intramolecular silylformylation reaction of acetylenes was reported by Alper and Matsuda in 1995 [47]. Pent-4-ynylmethylphenylsilane (Scheme 3, n = 1, R^1 = Me, R^2 = Ph, R^3 = H) was initially treated with triethylamine (1.0 equiv.), a catalytic amount of a rhodium catalyst under CO atmosphere (20 atm) and quite mild experimental conditions (40 °C, 24 h). Both the zwitterionic complexes $(\eta^6\text{-}C_6H_6BPh_3)^-Rh^+(1,5\text{-COD})$ (Rh^{sw}) and $Rh_4(CO)_{12}$ were effective, giving the corresponding aldehyde in good yields. According to Baldwin's rules [48,49], only the *exo-dig* cyclization occurred, generating 2-(formylmethylene)-1-silacycloalkanes with complete regio and stereoselectivity. None of the products derived from an *endo-dig*-mode cyclization or an intermolecular silylformylation were produced.

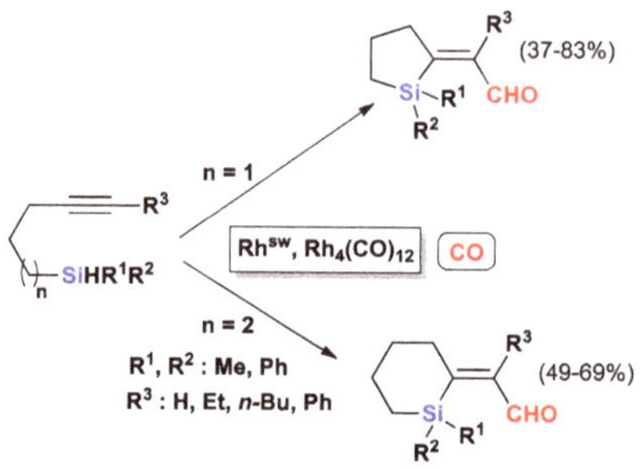

Scheme 3. First example of intramolecular silylformylation reaction of acetylenes reported by Alper and Matsuda.

The same trend was observed in the case of hexynylsilanes (Scheme 3, n = 2), which afforded the corresponding six ring silacycloalkanes. The yield of aldehyde was affected by the nature of R^1R^2HSi- group connected to the alkyl chain: higher product amounts were isolated when alkynylmethylphenylsilanes (Scheme 3, R^1 = Me, R^2 = Ph) were reacted rather than alkynyldiphenylsilanes. The intramolecular silylformylation proceeded smoothly also for internal alkynylsilanes (Scheme 3, R^3 = Et, n-Bu, Ph) regardless of the alkyl and aryl substituent. Thus, this method provides a vehicle for complete regio- and stereoselective formylation of acetylenic bonds.

The general mechanism proposed by the authors (Scheme 4, X = CH_2) involved initially an oxidative addition of Si-H to the rhodium catalyst, *cis* addition of the Rh-Si species to the triple bond followed by CO insertion into the Rh—C bond and reductive elimination with regeneration of the catalyst and formation of the -CHO group.

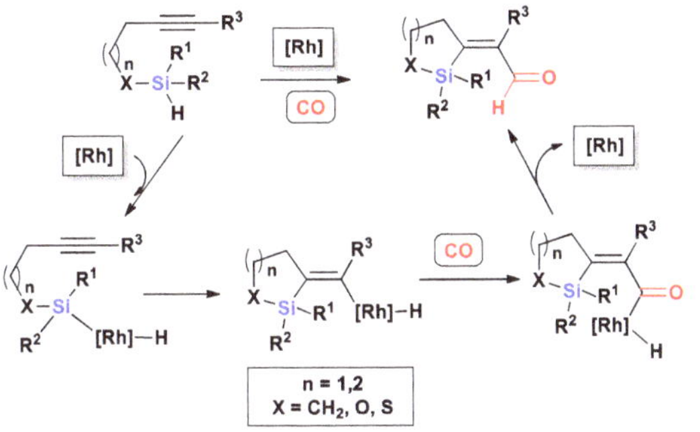

Scheme 4. General mechanism for intramolecular silylformylation reactions.

A few years later, Aronica and co-workers investigated the reactivity of both linear and C_3-branched 6-(methylphenylsilyl)-1-hexynes [50]. Linear substrate was first tested in the intramolecular silylformylation reaction, promoted by both zwitterionic Rh^{sw} and covalent complexes such as $Rh(acac)(CO)_2$ and $Rh_4(CO)_{12}$. In all cases, pure aldehyde was obtained in good yields with complete regioselectivity, i.e., exclusive addition of the -CHO moiety to terminal *sp*-carbon atom (Scheme 5).

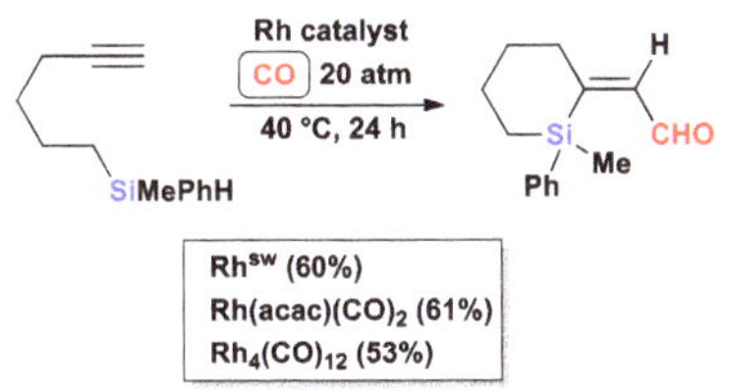

Scheme 5. Intramolecular silylformylation reaction of linear ω-silylacetylenes.

As a consequence, the air-stable Rh^{sw} species was used in subsequent reactions of C_3-branched acetylenes. As is evident from Scheme 6, the presence of an -R group did not influence the regioselectivity of the process, which afforded exocyclic isomers exclusively. On the other hand, when a bulky substituent, such as a *tert*-butyl group, was bonded to the alkyl chain, higher CO pressure (50 atm) and longer reaction times (48 h) were required to improve the yield of the silacyclane. One of the most interesting features of these reactions concerns the stereoselectivity: the presence of two chiral centers (i.e., Si* and C*-R) involved the possible formation of two different diastereomers, *cis* and *trans*. Unexpectedly, if a mixture of both isomers was obtained for the intramolecular silylformylation of 3-methyl-6-(methylphenylsilyl)-1-hexyne (Scheme 6, R = Me), the cyclization involving the *tert*-butyl derivative (Scheme 6, R = *t*-Bu) afforded the *trans* product exclusively.

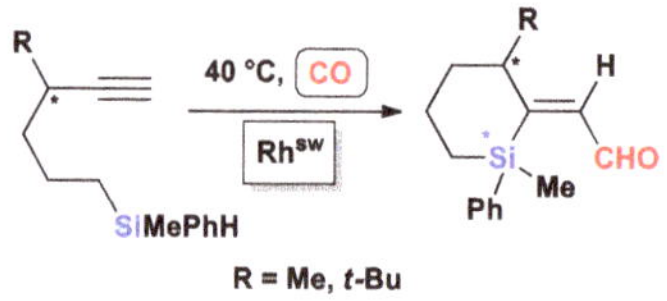

Scheme 6. Intramolecular silylformylation reaction of branched ω-silylacetylenes.

The same result was obtained [50] performing the reaction with a different catalyst, the co-condensate Rh/mesitylene, prepared according to the MVS technique [35,51–53] and consisting in a solution of small Rh metal clusters. The MVS species revealed a catalytic activity comparable with that of conventional organometallic compounds, high regio and disterostereoselectivity being observed also in this case.

Silacycloalkanes have been investigated as new and promising pharmaceutical substances [54]. Some of them have been tested as agents acting on the nervous system and showed activity as antitremorine compounds and gave promising results in the treatment of depression. Moreover, silicon derivatives were also proposed for the treatment or prevention of psoriasis and panic disorder. Some silacyclic derivatives exhibited high cytotoxicity and a broad spectrum of fungicidal activity. Finally, silacyclane compounds have been investigated as odorants since they showed quite different olfactory properties with respect to their carbon analogs, thus opening new possibilities for the fragrance industry.

2.2. The Intramolecular Silylformylation of ω-Bis(Dimethylsilylamino)Alkynes: Synthesis of Azasilacyclanes

The only example of intramolecular silylformylation of dimethylsilylaminoacetylenes was reported by Ojima and Vidal [55]. As a model reaction they reacted 1-bis(dimethylsilylamino)-3-octyne with CO in the presence of three different Rh-Co catalysts (Scheme 7). Unfortunately, the obtained azasilacyclopentane was highly unstable. Nevertheless, the authors observed that a stable product was generated by removing the silyl group connected to the nitrogen atom with $NaBH_4$ with contemporary reduction of the -CHO moiety.

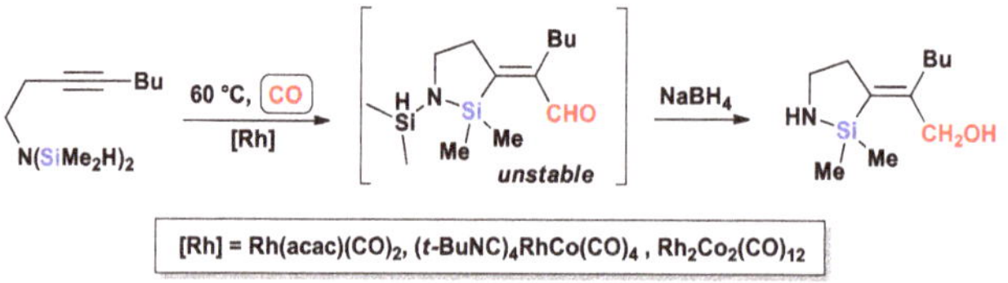

Scheme 7. Intramolecular silylformylation/desilylation of 1-bis(dimethylsilylamino)-3-octyne.

Thus coupling the silylformylation together with the reduction/desilylation step, azasilacyclopentane and cyclohexane were obtained in high yields, as depicted in Scheme 8.

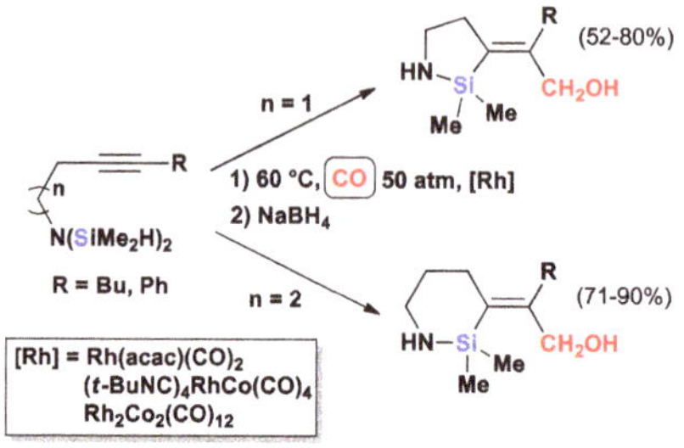

Scheme 8. Intramolecular silylformylation/desilylation reaction of *bis*(silyl)amino-alkynes.

Similar experimental conditions (60 °C, 10 atm CO, 14 h, Rh(acac)(CO)$_2$, (t-BuNC)$_4$RhCo(CO)$_4$ or Rh$_2$Co$_2$(CO)$_{12}$, then NaBH$_4$) were also applied to the intramolecular silylformylation/desilylation of (dimethylsilylamino)hexynylcyclohexane and cyclopentane derivatives which, after treatment with NaBH$_4$, gave the corresponding azasilabicycloalkenes (Scheme 9).

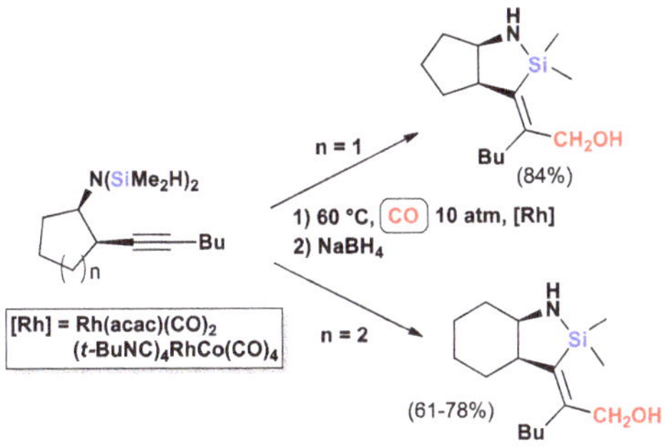

Scheme 9. Intramolecular silylformylation reaction of (dimethylsilylamino)hexynylcycloalkanes.

Surprisingly, when a dimethylsilyl group was replaced by a diphenylsilyl group, intramolecular silylformylation afforded the corresponding silapiperidine product in 63% isolated yield (Scheme 10).

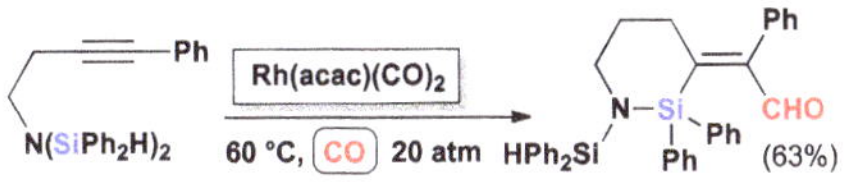

Scheme 10. Synthesis of 1,1-diphenyl-2-silyl-6-(1-formyl-1-benzylidene)azasilacyclohexane via intramolecular silylformylation reaction.

2.3. The Intramolecular Silylformylation of ω-Silyloxyalkynes: Synthesis of Oxasilacyclanes

In 1995, Ojima and co-workers described the first case of intramolecular silylformylation of terminal and internal alkynes featured by a dimethylsiloxy moiety as a directing group [56]. In agreement with the intramolecular silylformylation of ω-silylacetylenes, complete regio- and stereoselectivity was observed. Cyclization reactions of ω-(dimethylsiloxy)-alkynes were carried out in the presence of (t-BuNC)$_4$RhCo(CO)$_4$, Rh$_2$Co$_2$(CO)$_{12}$ or Rh(acac)(CO)$_2$ as catalyst, in toluene at 60–70 °C for 3–14 h to give the corresponding 3-exo-(1-formylalkylidene)oxasilacycloalkanes (Scheme 11). Both oxa-silacyclopentanes (Scheme 11, n = 1) and oxa-silacyclohexanes (Scheme 11, n = 2) were achieved in good yields regardless of the nature of the catalyst.

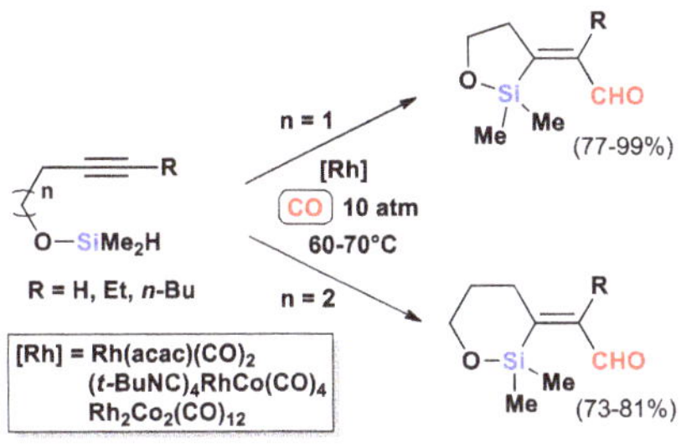

Scheme 11. First example of intramolecular silylformylation reaction of ω-(dimethylsiloxy)alkynes.

The intramolecular silylformylation was applicable also to cyclic systems. Indeed, the same authors tested the reactivity of O-(dimethylsilyl)-2-ethynyl and 2-propynyl derivatives depicted in

Scheme 12 [56]. All reactions proceeded smoothly at 65 °C and 10 atm CO, with (*t*-BuNC)$_4$RhCo(CO)$_4$ as the best catalyst.

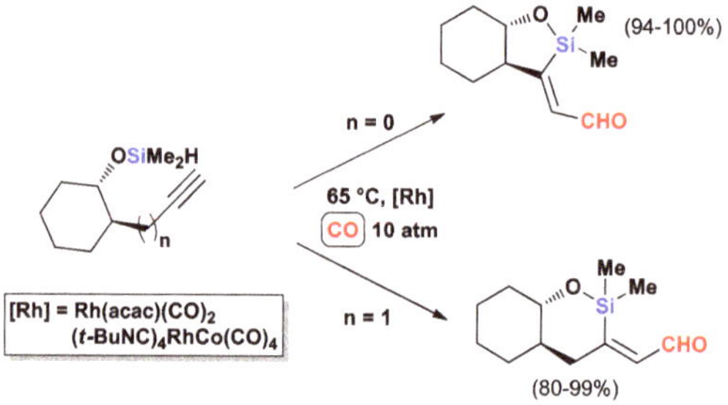

Scheme 12. Intramolecular silylformylation of cyclic *O*-(dimethylsilyl)-2-ethynyl and propynyl derivatives.

As solvated metal atoms prepared according to the MVS technique had revealed high reactivity and selectivity in silylformylation reactions [35], it was interesting to verify the catalytic activity of these species in the intramolecular processes of 2-(dimethylsiloxy)-4-nonyne derived from a homopropargyl alcohol. Complete regio/stereoselectivity was observed: 3-(1′-formylpentylidene)-1-oxa-2-silacyclopentane was obtained in high yield (86%) (Scheme 13) [57].

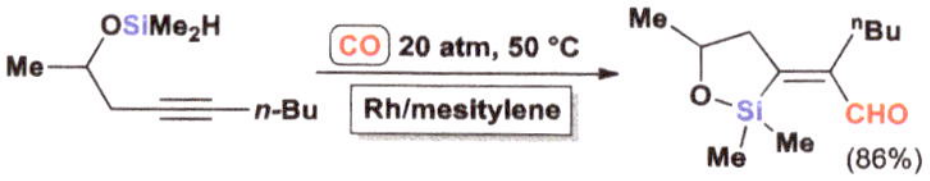

Scheme 13. Intramolecular silylformylation of 2-(dimethylsiloxy)-4-nonyne promoted by a Rh/mesitylene obtained via the Metal Vapor Synthesis (MVS) technique.

A few years later, starting from dimethylsiloxyalkadiynes, Bonafoux and Ojima developed a process of desymmetrization based on a single intramolecular silylformylation reaction [58]. As described in Scheme 14, Rh(acac)(CO)$_2$ was effective in promoting the reactions of terminal and internal alkynes at room temperature and under 10 atm of carbon monoxide. Both cyclizations took place smoothly but 5-exo(formylmethylene)oxacyclopentane (Scheme 14, R = H) could not be purified as it decomposed when subjected to silica gel chromatography. Reduction of the formyl moiety with NaBH$_4$ afforded the corresponding alcohol, isolated in good yield (70%). The highly functionalized cyclic products thus obtained represent useful synthetic intermediates, since they can be manipulated at the unreacted acetylene moiety as well as at the -CHO or -CH$_2$OH functional groups.

Scheme 14. Desymmetrization of dimethylsiloxyalkadiynes based on intramolecular silylformylation reaction.

The same authors also described a three-steps protocol for the synthesis of 5-(2-acetoxyalkyl)-2-oxa-1-silacyclopentenes [59]. The sequence started with the intramolecular silylformylation of ω-(dimethylsiloxy) alkynes, followed by reduction of the corresponding aldehyde to give 5-exo-(hydroxyethylene)-2-oxa- 1-silacyclopentanes. Subsequent DMAP-catalyzed treatment of the obtained alcohols with acetic anhydride involved a skeletal rearrangement which afforded the corresponding oxasilacyclopentenes exclusively (Scheme 15).

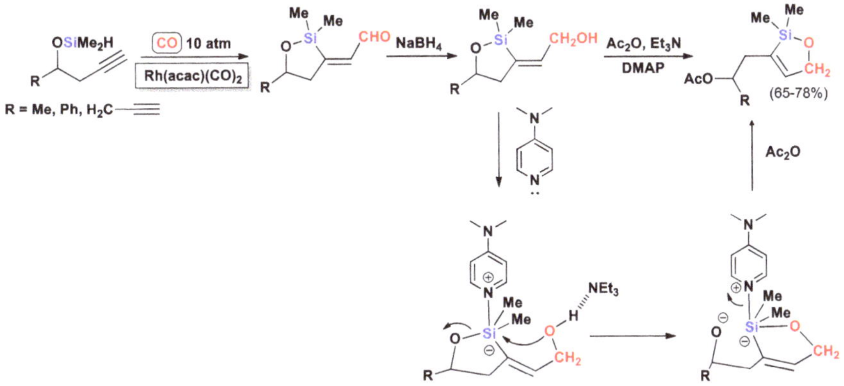

Scheme 15. Three-steps protocol for the synthesis of 5-(2-acetoxyalkyl)-2-oxa-1-silacyclopentenes.

Moreover, the authors observed that when O-dimethylsilylethynylcyclohexanol was submitted to the same transformations, (2-(2,2-dimethyl-2,5-dihydro-1,2-oxasilol-3-yl)cyclohexyl acetate was isolated in 76% yield (Scheme 16). The rearrangement products containing an acylated moiety together with an oxasilacyclopentene nucleus could be employed as useful polyfunctionalized intermediates in organic chemistry.

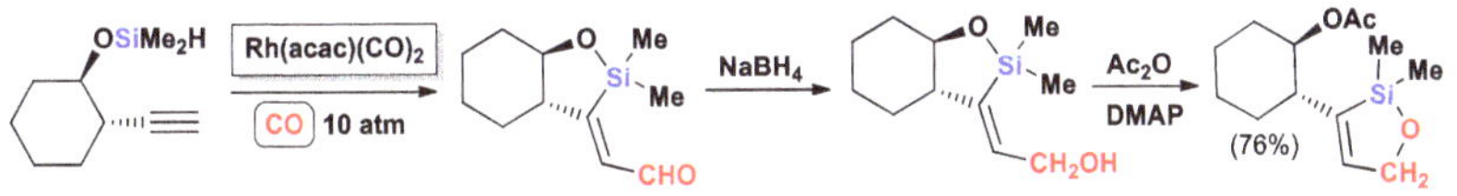

Scheme 16. Skeletal rearrangement of O-dimethylsilylethynylcyclohexanol.

Another interesting application of oxa-silacyclopentanes was reported in 2003 by Denmark and Kobayashi [60]. First, intramolecular silylformylation of alkynyloxyhydrosilanes was carried out under CO pressure (10 atm) at 70 °C. (t-BuNC)$_4$RhCo(CO)$_4$ showed the best catalytic efficiency, affording the five-membered cyclic silyl ethers in 72% yield. (Scheme 17, step 1). With the (1-formylalkylidene)oxa-silacycloalkanes in hands, authors investigated the possible cross-coupling of heterocyclic compounds. Initially, a deep investigation on the experimental conditions was performed: DMF resulted as the best solvent, the combination of [(allyl)PdCl]$_2$ and CuI the optimal catalytic species and KF was chosen as a fluoride source. Then, oxa-silacyclopentanes were reacted with several aromatic iodides affording the corresponding α,β-unsaturated aldehydes (Scheme 17, step 2). Electrophiles with electron donating groups reacted more slowly than those bearing electron-withdrawing moieties, and the reaction of the cyclic silylether possessing a methyl group on the alkene was slower than the reaction of the terminal derivative. Nevertheless, cross-coupling products were achieved in good to excellent yields (57–93%).

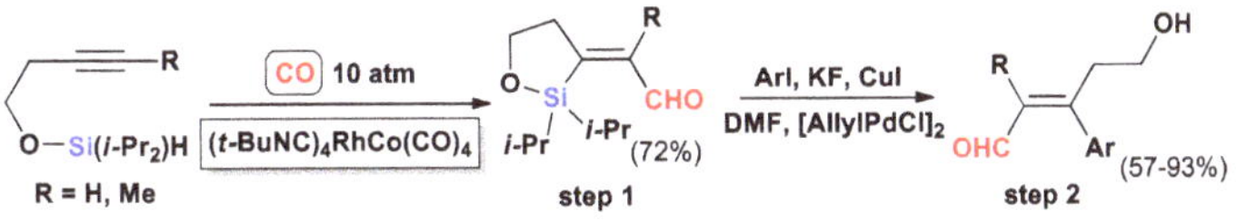

Scheme 17. Tandem intramolecular silylformylation/cross-coupling reactions of alkynyloxyhydrosilanes.

Finally, Leighton and co-workers developed several sequential approaches to polyol derivatives via oxa-silacyclanes intermediates, which were generated in situ and then converted into polyketides fragments by means of Tamao oxidations (Scheme 18) [61–67].

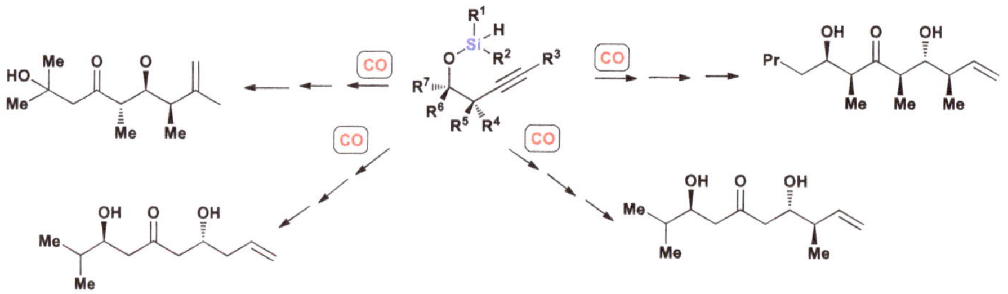

Scheme 18. Tandem intramolecular silylformylation/crotylsilylation reactions.

3. Heterocycles Synthesis via Metal-Catalyzed Silylcarbocyclization of Alkynes

As well shown in the previous section, intramolecular hydrosilylation and silylformylation reactions of alkynes represent a valid route to several types of highly functionalized silacycles. Silylcarbocyclization (SiCAC) protocols are instead a different synthetic approach to heterocyclic compounds: these transition metal-catalyzed tandem addition/cyclization reactions of alkynes with hydrosilanes, often performed under carbonylative atmosphere, are very useful for obtaining highly functionalized heterocycles bearing exocyclic silyl moieties, sometimes amenable for further *one-pot* synthetic transformations.

The first silylcarbocyclization was serendipitously discovered by Ojima et al. in 1991: during their studies on the silylformylation of 1-hexyne with triethylsilane, carried out in the presence of $Co_2Rh_2(CO)_{12}$ as catalyst, in addition to the usual hydrosilylation and silylformylation products they observed a small amount of 2,4-dibutyl-3-(triethylsilyl)-cyclopent-2-en-1-one (Scheme 19) [25]. However, only mechanistic studies performed in a following paper better clarified the origin of this cyclic product: a metal-promoted silylcarbonylation of 1-hexyne gave the β-silylacryloyl-metal intermediate (**I**), which in turn then provided a carbometalation on a second 1-hexyne molecule to generate (**II**); after the following carbocyclization and β-hydride elimination steps, a highly regioselective reduction of intermediate (**IV**) took place at the less sterically hindered double bond; finally, a hydrogen-metal exchange between species (**V**) and a further triethylsilane molecule gave the final cyclopentenone product (Scheme 19) [26].

Several synthetic applications of silylcarbocyclization reactions have been previously treated as part of more general reviews, focused on the transition metal-promoted cyclizations [68–70] or on the chemistry of hydrosilanes with alkynes [27]; however, a complete and up-to-date overview of SiCAC protocols for the preparation of heterocycles is still missing. Therefore, in the second part of the present review we shall try to provide an exhaustive and critical account of this literature, giving special emphasis on the adopted catalytic systems.

Silylcarbocyclization reactions applied to the synthesis of heterocyclic compounds can be divided into three main groups, depending on the starting alkynes: (i) *standard* silylcarbocyclizations, mainly involving allyl propargyl and dipropargyl ethers/amines, which gave in most cases tetrahydrofuran and pyrrolidine derivatives (Scheme 20, path a); (ii) *cascade* silylcarbocyclizations, involving instead enediynes and triynes with a suitable chemical structure, which led to the formation of fused tricyclic structures (Scheme 20, path b); (iii) *heteroatom-promoted* silylcarbocyclizations, involving ethynyl alcohols and amines, where lactones and lactams were obtained (Scheme 20, path c). The literature will be organized below following this systematic approach.

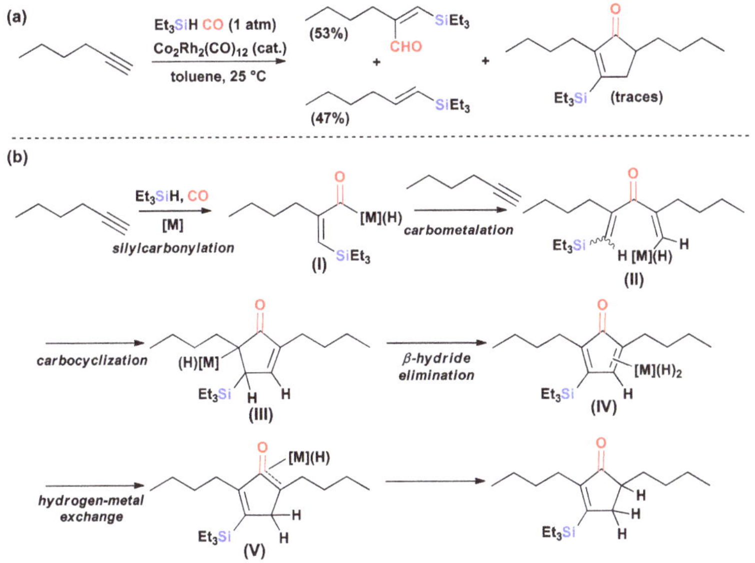

Scheme 19. First example of silylcarbocyclization reaction reported by Ojima and co-workers (**a**) and related reaction mechanism (**b**).

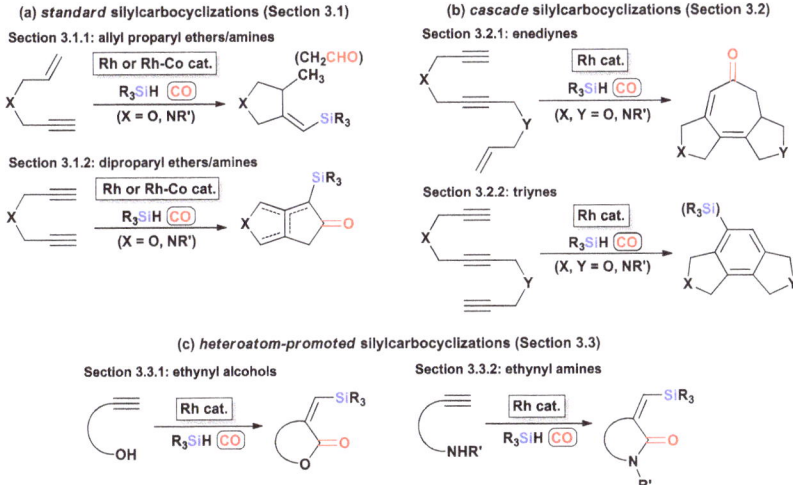

Scheme 20. Classification of SiCAC reactions: (**a**) *standard* silylcarbocyclizations, involving allyl propargyl and dipropargyl ethers/amines (Section 3.1); (**b**) *cascade* silylcarbocyclizations, involving enediynes and triynes (Section 3.2); (**c**) *heteroatom-promoted* silylcarbocyclizations, involving ethynyl alcohols/amines (Section 3.3).

3.1. Synthesis of Heterocycles via Metal-Catalyzed Standard Silylcarbocyclizations of Alkynes

Transition metal-catalyzed *standard* silylcarbocyclizations of alkynes represent the most common synthetic approach to heterocycles based on SiCAC protocols and involve two main classes of substrates: enynes (i.e., allyl propargyl ethers/amines) and diynes (i.e., dipropargyl ethers/amines). In particular, SiCAC of allyl propargyl ethers and amines allow the formation of 3-((triorganosilyl)methylene)tetrahydrofuran and 3-((triorganosilyl)methylene)pyrrolidine scaffolds, respectively; instead, SiCAC of dipropargyl ethers and amines often lead to more complicated cyclopentafuranone and cyclopentapyrrolone derivatives, although in some cases functionalized tetrahydrofurans/pyrrolidines or piperidinones were also obtained. In general, *standard* SiCACs proceed successfully with both alkyl and aryl silanes, often performed under CO (at atmospheric or high pressure) and using rhodium or rhodium-cobalt complexes as catalysts.

3.1.1. Standard Silylcarbocyclizations of Allyl Propargyl Ethers/Amines

The first investigation on *standard* silylcarbocyclization of allyl propargyl ethers and amines was reported in 1992 by Ojima and co-workers [71]. They described the reaction of allyl propargyl ether with dimethylphenylsilane, performed in toluene at 70 °C and under CO pressure (1 atm), in the presence of $Rh_4(CO)_{12}$ as catalyst: 3-(silylmethylene)-4-methyltetrahydrofuran was obtained in 61% yield after 18 h. Interestingly, the same product was obtained in higher yields (85%) using $Rh(acac)(CO)_2$ as the catalytic system and under N_2 atmosphere, thus demonstrating that *standard* SiCAC reactions do not strictly require carbon monoxide. A three-step mechanism was hypothesized for this transformation, consisting of silylmetalation of the triple bond, carbocyclization and H-shift (Scheme 21). In the same paper, the authors also described a similar SiCAC reaction for diallyl propargyl amine with $PhMe_2SiH$ ($Rh(acac)(CO)_2$, CO 1 atm, 70 °C), which gave the corresponding pyrrolidine as the only product in almost quantitative yield.

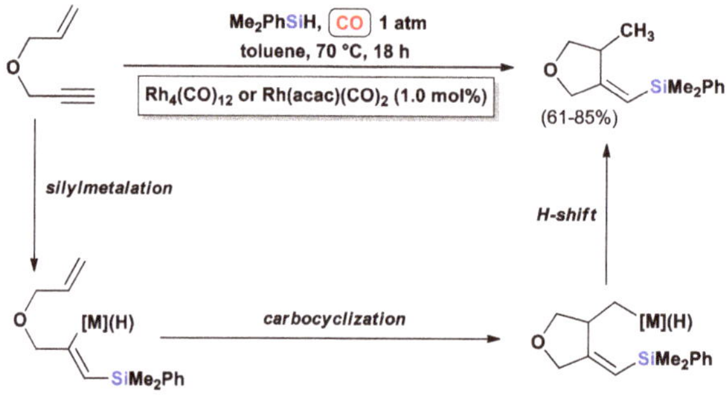

Scheme 21. SiCAC of allyl propargyl ether with Me_2PhSiH: 3-(silylmethylene)-4- ethyltetrahydrofuran was obtained through a three-steps mechanism, i.e., silylmetalation, carbocyclization and H-shift.

In a following paper, the same group extended *standard* SiCAC to a more structurally complex allyl propargyl ether [72]. Working under the same conditions of their previous work (1.0 equiv. of Me_2PhSiH, 1 mol% of $Rh(acac)(CO)_2$, 1 atm of CO, at 50 °C in toluene), the expected SiCAC product was recovered as a mixture with the corresponding carbonylative SiCAC (namely, CO-SiCAC) derivative, arising from a carbon monoxide insertion between the carbocyclization and H-shift steps. Interestingly, a fine tuning of the experimental conditions may influence the product selectivity: when the reaction was run in *n*-hexane 1 M under 1 atm of CO, the only SiCAC product was isolated in 60% yield; instead, in THF 0.07 M with 2.6 atm of CO, the CO-SiCAC product was found the most predominant compound in 95% yield (Scheme 22).

In 2002 Ojima et al. reported a more detailed investigation on Rh-catalyzed SiCAC of enynes, with special focus on allyl propargyl amines for the synthesis of pyrrolidine derivatives [73]. Analogous to their previous work on hepta-1,6-dien-4-yl propargyl ether [72], the selectivity toward SiCAC or CO-SiCAC products can be controlled depending on the experimental conditions, while always working with 0.5 mol% of $Rh_4(CO)_{12}$ as catalyst: with an excess (1.5 equiv.) of hydrosilane at 0.4 M concentration in *n*-hexane, at 22 °C under atmospheric pressure of CO, the corresponding SiCAC products were surprisingly obtained in less than 1 min (yields: 74–89%).

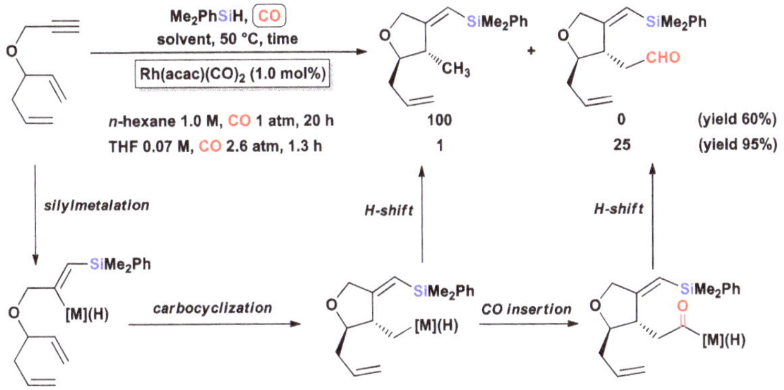

Scheme 22. *Standard* silylcarbocyclization of hepta-1,6-dien-4-yl propargyl ether with Me$_2$PhSiH.

The exclusive formation of CO-SiCAC pyrrolidines (56–85% yields) was instead found by using an almost equimolar amount of silane (1.05 equiv.) at 0.02 M concentration in 1,4-dioxane, under 20 atm of CO at 105 °C, in the presence of 10 mol% of P(OEt)$_3$ as ligand (Scheme 23). Since high reactants dilution is not advantageous in organic synthesis, authors also investigated a further optimization of the protocol for obtaining CO-SiCaC products: the amine solution in 1,4-dioxane was cooled before the addition of Rh$_4$(CO)$_{12}$, hydrosilane and P(OEt)$_3$; then, the frozen reaction mixture was placed in autoclave and pressurized with CO (20 atm). This "freeze and CO" protocol was able to block the SiCaC reaction by freezing the reaction to start until the whole system is under high carbon monoxide pressure, thus favoring the formation of the CO-SiCaC product.

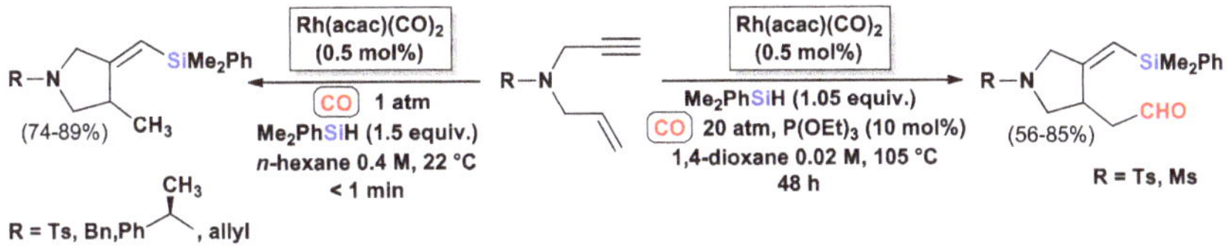

Scheme 23. Synthesis of pyrrolidine derivatives via *standard* silylcarbocyclization of allyl propargyl amines.

Matsuda et al. also reported an interesting study on rhodium-catalyzed *standard* silylcarbocyclization of 1,6-enynes derivatives, including allyl propargyl ethers and amines [74]. As previously observed by Ojima and coll. [72,73], working with Rh$_4$(CO)$_{12}$ or Rh(acac)(CO)$_2$ catalysts under high CO pressure (20 kg/cm^2) they usually obtained the selective formation of the CO-SiCAC product; however, in the case of allyl propargyl benzylamine the SiCAC pyrrolidine compound was obtained as the sole product under the same experimental conditions. Although the authors did not provide any explanation for this result, they believe that the role of benzyl substituent on the nitrogen atom is crucial for explaining this different reactivity.

In 2003, Chung and collaborators proposed the first application of a supported and recoverable catalyst in silylcarbocyclization reactions, i.e., bimetallic Co/Rh nanoparticles immobilized on charcoal [75]. The supported catalyst was very easily prepared by refluxing Co$_2$Rh$_2$(CO)$_{12}$ with charcoal in THF, giving the final material with a fixed 2:2 cobalt-rhodium stoichiometry. It was then successfully employed in the *standard* SiCAC of a large family of 1,6-enynes, including allyl propargyl ethers bearing internal acetylene and/or alkene moieties: in particular, working with a large excess (5.0 equiv.) of hydrosilane at 105 °C in 1,4-dioxane under atmospheric pressure of CO, the corresponding CO-SiCAC products were obtained (23–87% yields) after 12 h; instead, SiCAC THF derivatives were recovered in satisfactory yields after only 2 h by using the same excess of hydrosilane, in *n*-hexane as solvent at 22 °C and without carbon monoxide atmosphere (Scheme 24). Therefore, the present *standard* silylcarbocyclization protocol appears quite interesting, both for milder reaction conditions of CO-SiCAC pathways (often requiring high CO pressure and reagents concentration) and for catalyst recyclability.

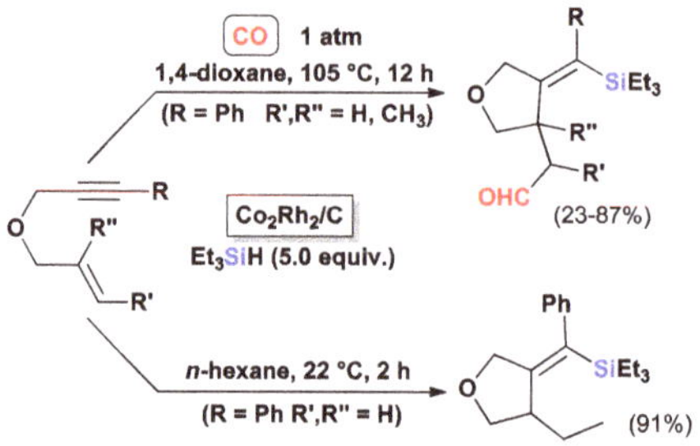

Scheme 24. *Standard* silylcarbocyclization of allyl propargyl ethers bearing internal acetylene and/or alkene moieties, catalyzed by Co/Rh nanoparticles immobilized on charcoal.

In the last fifteen years, *standard* silylcarbocyclization protocols have been mostly applied as a key step for the synthesis of more complex heterocyclic compounds, including biologically active products. In 2006, Murai et al. reported an extended investigation on the synthesis of 1,2,2′-trisubstituted pyrrolidines and piperidines through the reaction of thioiminium salts (derived from the addition of lithium acetylides to γ-and δ-thiolactams) with proper Grignard reagents [76]. In order to show the synthetic applicability of the obtained compounds, N-allyl-2-ethynyl-2-substituted pyrrolidines and piperidines were then also subjected to *standard* silylcarbocyclization: reactions were performed with $Rh_4(CO)_{12}$ (0.5 mol%) as catalyst and Me_2PhSiH (1.5 equiv.) as silane, in *n*-hexane under CO atmosphere and at room temperature. In the case of pyrrolidines, SiCAC afforded 1,2,7a-trisubstituted hexahydro-1H-pyrrolizines as a mixture of four diastereomers, the stereochemistry of which was identified by NOESY spectroscopy; instead, *standard* SiCAC of piperidine reagents gave 1,2,8a-trisubstituted octahydroindolizines as a mixture of only two diastereomers (Scheme 25).

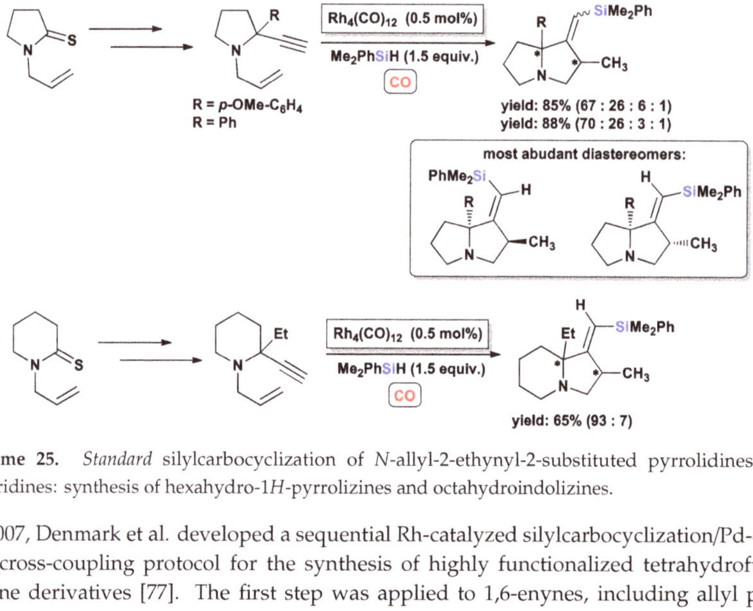

Scheme 25. *Standard* silylcarbocyclization of N-allyl-2-ethynyl-2-substituted pyrrolidines and piperidines: synthesis of hexahydro-1H-pyrrolizines and octahydroindolizines.

In 2007, Denmark et al. developed a sequential Rh-catalyzed silylcarbocyclization/Pd-catalyzed Hiyama cross-coupling protocol for the synthesis of highly functionalized tetrahydrofuran and pyrrolidine derivatives [77]. The first step was applied to 1,6-enynes, including allyl propargyl ethers and amines, under typical SiCAC conditions: $Rh_4(CO)_{12}$ (0.5–5 mol%) as catalyst, an excess

(1.5–2.0 equiv.) of silane, under atmospheric CO pressure at the selected temperature. Interestingly, the triorganosilyl moiety of the corresponding heterocyclic products can be then subjected to a Hiyama cross-coupling reaction with aryl iodides in the presence of a suitable palladium catalyst. The best performance was observed with 2.5 mol% of Pd$_2$(dba)$_3$·CHCl$_3$ and 2.0 equiv. of TBAF·3 H$_2$O as additive, in THF at room temperature: both electron rich and electron poor aryl iodides gave the coupling products in good yields (Scheme 26). More recently, the same research group applied *standard* silylcarbocyclization to a highly functionalized allyl propargyl amine, i.e., N-tosyl-N-methylpropargyl-(L)-vinylglycine methyl ester, as a key step for the total synthesis of isodomoic acids G and H, two kainoid amino acid derivatives isolated from red alga *Chondria armata* with well recognized properties as neuroexcitatory agents [78,79].

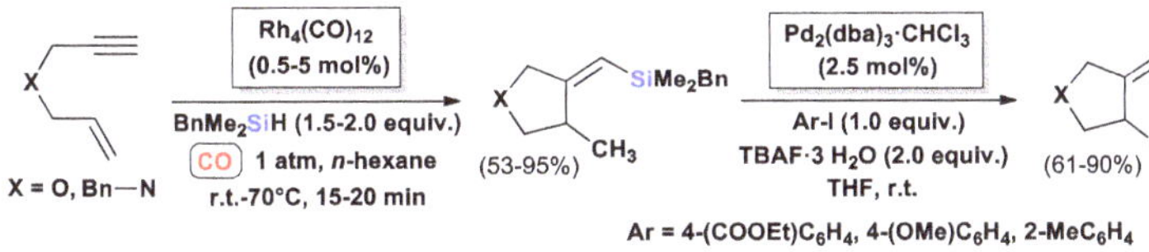

Scheme 26. Sequential Rh-catalyzed *standard* silylcarbocyclization/Pd-catalyzed Hiyama cross-coupling of allyl propargyl ethers and amines.

3.1.2. Standard Silylcarbocyclizations of Dipropargyl Ethers/Amines

Mostly investigated by the Ojima's group, *standard* silylcarbocyclizations of dipropargyl ethers and amines represent a useful and rapid tool for obtaining O- and N-containing bicyclic compounds (especially cyclopentafuranone and cyclopentapyrrolone derivatives).

In 1992, they reported the first attempt of SiCAC on allyldipropargylamine as starting substrate [71]. The reaction was performed with HSiEt$_3$ (3.0 equiv.) in the presence of bimetallic (t-BuNC)$_4$RhCo(CO)$_4$ (0.25 mol%) as catalyst, in toluene at 65 °C and 50 atm of carbon monoxide: after 48 h, a bicyclic compound incorporating two CO units was obtained as a predominant product (62% yield), together with a small amount (<2% yield) of a piperidone derivative arising from a single CO incorporation. Interestingly, when the same reaction was performed with the more common Rh$_4$(CO)$_{12}$ catalyst and under atmospheric CO, the same piperidone was found as the sole product in good yields (81%). The proposed mechanism involved the silylmetalation of an alkynyl moiety of starting allyldipropargylamine, followed by CO insertion and carbocyclization steps; the obtained intermediate may then follow two different pathways: (i) second CO insertion and carbocyclization steps, followed by β-hydride elimination of [M]H, regioselective addition of [M]H and regeneration of Et$_3$Si[M], affording a final bicyclic SiCAC product; (ii) hydrosilylation of a second molecule of silane, followed by Et$_3$Si[M] regeneration to give the final piperidone derivative (Scheme 27).

In a following paper, authors used very similar experimental conditions for the (t-BuNC)$_4$RhCo(CO)$_4$ or Co$_2$Rh$_2$(CO)$_{12}$ catalyzed SiCAC of benzyldipropargylamine with t-butyldimethylsilane [80]: surprisingly, 7-azabicyclo[3.3.0]oct-1-ene was found (60% yield), together with small amounts of its $\Delta^{1,5}$-isomer, which can be easily converted into 7-azabicyclo[3.3.0]oct-1-ene by in situ treatment with RhCl$_3$·3 H$_2$O at 50 °C. A plausible mechanism involved the starting silylmetalation of a triple bond with t-BuMe$_2$SiH, followed by a sequence of carbocyclization, CO insertion and carbocyclization to give the bicyclic intermediate **A**, from which both final products can be obtained: (i) though a sequential β-hydride elimination, regioselective hydrometalation and β-hydride elimination, 7-azabicyclo[3.3.0]oct-1-ene was obtained; (ii) its Δ^5-isomer was instead obtained with a 1,3-[M] shift step, followed by the regeneration of Et3Si[M] (Scheme 28) [81].

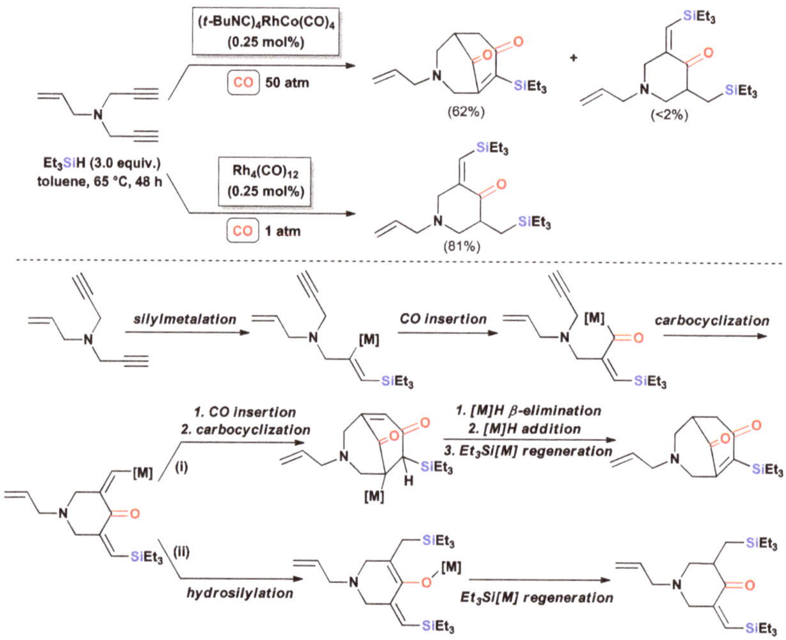

Scheme 27. First report of *standard* SiCAC of dipropargyl amines: two different mechanicistic pathways afforded a bicyclic compound incorporating two CO units (i) or a piperidone derivative incorporating a single CO unit (ii).

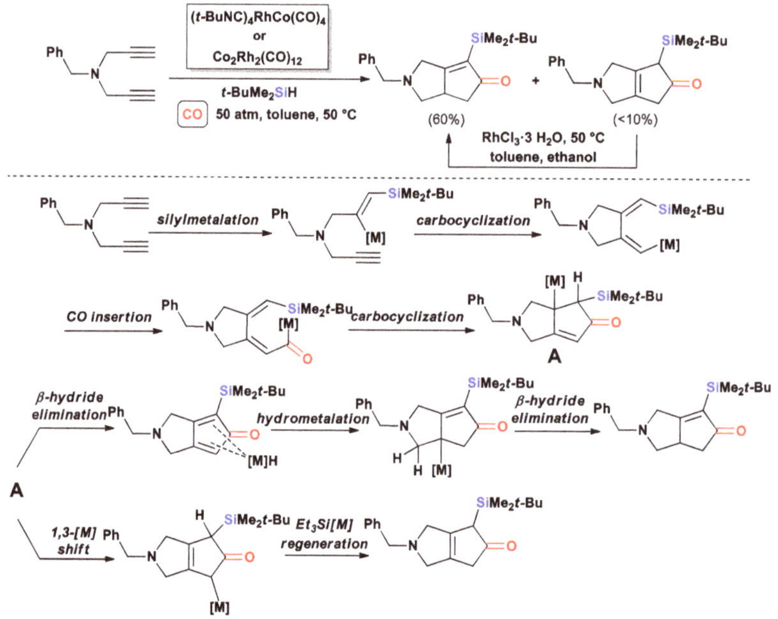

Scheme 28. *Standard* silylcarbocyclization of benzyldipropargylamine with *t*-butyldimethylsilane and proposed reaction mechanism.

However, a more extensive and detailed investigation on *standard* SiCAC of dipropargyl ether and amines was reported in 1998 by the same research group, performed by testing different substrates, Rh catalysts and experimental conditions [82]. Interestingly, they found that carbon monoxide pressure is a very critical parameter: working under high CO pressure (15–50 atm), SiCAC reactions proceeded as previously described [80,81] to afford heterobicyclo[3.3.0]octenones in good yields; instead, under ambient carbon monoxide pressure reactions occurred in a different way, affording tetrahydrofuran or pyrrolidine derivatives as final products (Scheme 29). In this last case, the CO insertion does not occur after the silylmetalation and carbocyclization steps, therefore a hydride shift and a subsequent 1,2- and/or 1,4-hydrosilylation can give final heterocyclic products.

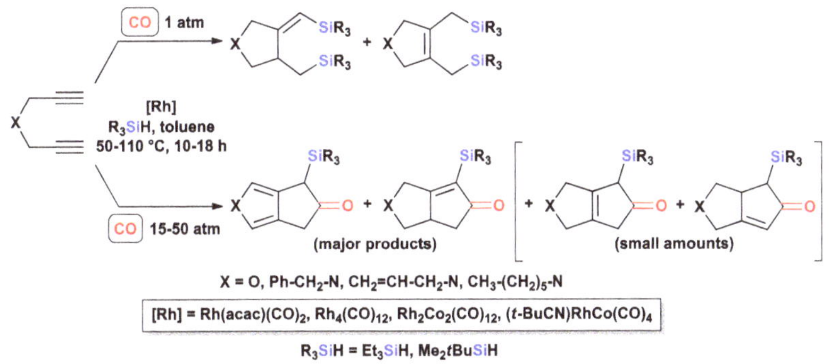

Scheme 29. Ojima's investigation on *standard* SiCAC of dipropargyl ether and amines.

In addition to the extensive studies performed by the Ojima's research group, *standard* SiCAC of dipropargyl ethers/amines were also investigated by Matsuda et al. [83]: reactions were performed with 0.5 mol% of $Rh_4(CO)_{12}$ as catalyst and $tBuMe_2SiH$ (2.0 equiv.) as silane, under high CO pressure (20 atm) at 95 °C and using benzene or CH_3CN as solvent, to give the corresponding heterobicyclo[3.3.0]octenones as a mixture of regioisomers.

To conclude this section on *standard* silylcarbocyclizations, it is worth spending a few words on allenynes, showing a reactivity very similar to diynes. In 2004, Shibata and co-workers studied rhodium catalyzed SiCAC of propargyl homoallenyl ethers and amines under atmospheric CO pressure, providing cyclic (tetrahydrofuran or pyrrolidine) 1,4-dienes [84]. Reactions proceeded smoothly on a wide range of substrates, with both trialkylsilanes and trialkoxysilanes, using $Rh(acac)CO_2$ complex (5 mol%) as the most efficient catalyst (Scheme 30). The proposed mechanism, supported by deuterium labeling experiment, involved a regioselective silylmetalation on the double bond of the allene moiety closer to the heteroatom, followed by carbometalation on the alkynyl group to give the corresponding cyclic vinyl rhodium complex; finally, reductive elimination provided the heterocyclic product with regeneration of the Rh catalyst.

3.2. Synthesis of Heterocycles via Metal-Catalyzed Cascade Silylcarbocyclizations of Alkynes

Transition metal-catalyzed *cascade* silylcarbocyclizations of alkynes have been less studied than *standard* SiCAC as they involved more complex substrates, i.e., enediynes and triynes with a suitable chemical structure. However, *cascade* SiCAC represent very elegant synthetic protocols for the selective synthesis of fused tricyclic structures: heteroatom congeners of hexahydro-1*H*-cyclopenta[*e*]azulen-5(6*H*)-one and hexahydro-*as*-indacene using, respectively, enediynes and triynes as starting alkynes. If *standard* SiCAC can be performed under atmospheric or high CO pressure (in few cases even without CO), all the reported *cascade* SiCAC protocols always used 1 atm of carbon monoxide. Concerning catalysts, rhodium or rhodium-cobalt complexes have been successfully tested also for these reactions.

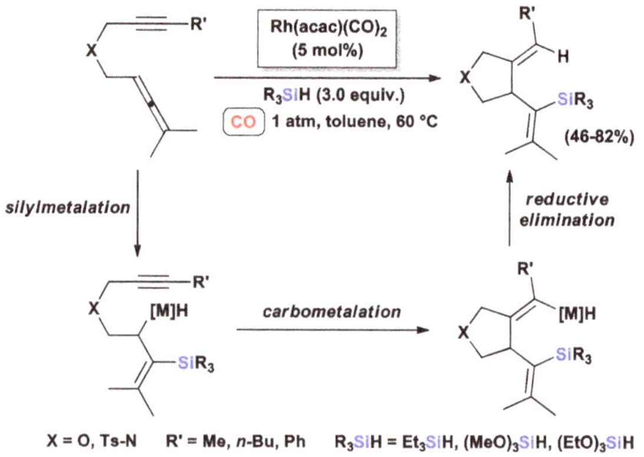

Scheme 30. *Standard* SiCAC of proparyl homoallenyl ethers and amines under atmospheric CO pressure: synthesis of cyclic (tetrahydrofuran or pyrrolidine) 1,4-dienes.

3.2.1. Cascade Silylcarbocyclizations of Enediynes

The first study on *cascade* silylcarbocyclization reactions of enediynes was reported in 2000 by Ojima's research group [85]. Starting from the excellent results of their previous investigations on *standard* SiCAC of enynes and diynes, they tried to extend a similar protocol to enediynes. Interestingly, when dodec-11-ene-1,6-diyne was treated with PhMe$_2$SiH (2.0 equiv.) in the presence of Rh(acac)(CO)$_2$ (1 mol%), at 70 °C in toluene as solvent and under atmospheric CO, hexahydro-1*H*-cyclopenta[*e*]azulen-5(6*H*)-one was obtained as the main product after only 1 h, together with small amounts of two bis(cyclopentylidene) derivatives. However, a fine tuning of the experimental conditions allowed to improve selectivity: in fact, hexahydro-1*H*-cyclopenta[*e*]azulen-5(6*H*)-one was obtained as the only product when SiCAC reaction was performed in THF at lower reagents concentration and at room temperature. This optimized protocol was then applied to other enediynes, including their oxygen or nitrogen congeners, to give the corresponding *O*- or *N*-containing fused tricyclic structures. The proposed reaction mechanism provides three sequential carbocyclization steps (hence the name "*cascade* SiCAC"): after starting silylmetalation of the terminal alkyne moiety, the first carbocyclization took place; because of the steric hindrance between vinylsilane and vinyl-rhodium moieties in the resulting intermediate, it was then subjected to an isomerization via the "Ojima-Crabtree mechanism", followed by the second carbocyclization step; the subsequent CO insertion step gave an acyl-rhodium intermediate, which was then subjected to the last carbocyclization, and a β-silyl elimination step afforded the final tricyclic product (Scheme 31).

More recently, Ojima and co-workers extended their studies on the scope and limitation of *cascade* SiCAC to 1-substituted dodec-11-ene-1,6-diynes and their heteroatom congeners [86]. When 1-methyl substituted dodec-11-ene-1,6-diyne was treated with PhMe$_2$SiH (0.5 equiv.) in the presence of [Rh(COD)Cl]$_2$ or Rh(acac)(CO)$_2$ as catalyst (1 mol%), at 70 °C in toluene and under atmospheric CO pressure, they did not find the expected hexahydro-1*H*-cyclopenta[*e*]azulen-5(6*H*)-one but the corresponding 5-6-5 fused tricyclic compound (i.e., incorporating no CO unit) as the only product (70% yield in the case of [Rh(COD)Cl]$_2$; 96% yield by using Rh(acac)(CO)$_2$). However, the authors serendipitously discovered that working under similar conditions but in the absence of hydrosilane, the hexahydro-1*H*-cyclopenta[*e*]azulen-5(6*H*)-one was instead obtained in good yield (Scheme 32). This last Rh-catalyzed reaction in the absence of hydrosilane is actually an intramolecular [2 + 2 + 2 + 1] cycloaddition, occurring through a mechanism totally different from *cascade* SiCAC, although the same type of products is formed. We will not take into account this reaction, which is beyond the scope of the present review, but it is worth emphasizing that it has been successfully applied to several 1-substituted dodec-11-ene-1,6-diynes, including their oxygen and nitrogen congeners [87].

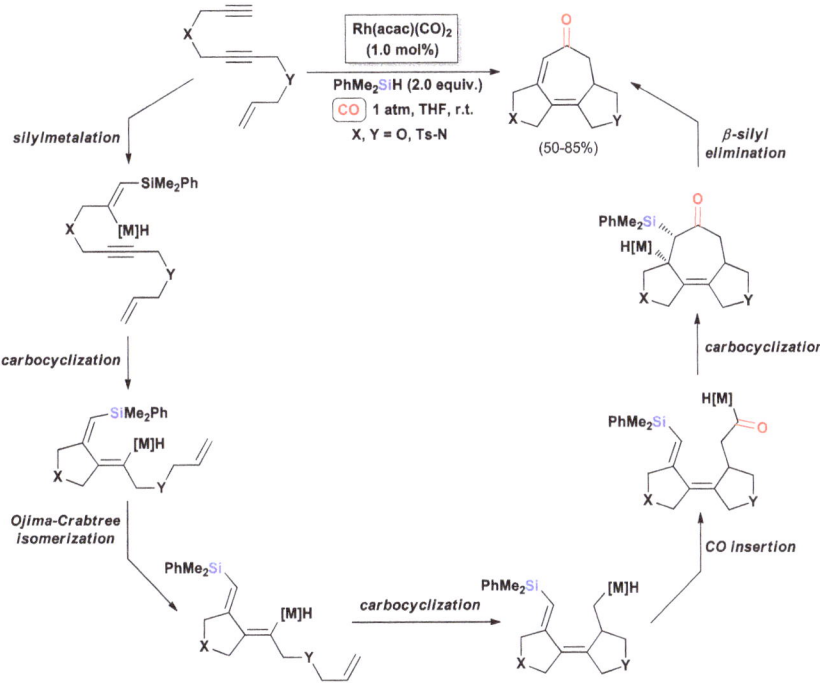

Scheme 31. First investigation of *cascade* SiCAC of enediynes: the proposed reaction mechanism involved three sequential carbocyclization steps, hence the name "cascade SiCAC".

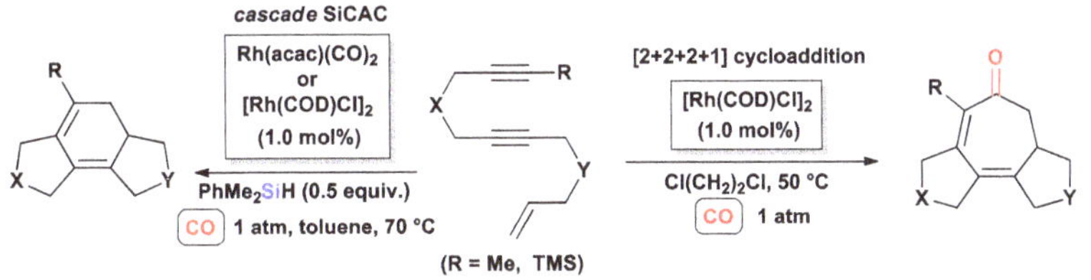

Scheme 32. Rhodium-catalyzed *cascade* SiCAC vs. intramolecular [2 + 2 + 2 + 1] cycloaddition of 1-substituted dodec-11-ene-1,6-diynes and their heteroatom congeners.

3.2.2. Cascade Silylcarbocyclizations of Triynes

Although less investigated than diynes, also triynes were successfully tested as starting substrates for *cascade* silylcarbocyclization reactions. In 1999, Ojima and collaborators treated dodec-1,6,11-triynes and some oxygen- and nitrogen-containing analogs with several hydrosilanes (1.0–2.0 equiv.), in toluene under atmospheric carbon monoxide pressure, using different Rh complexes (0.5–1.0 mol%) as catalyst, including $Rh_4(CO)_{12}$, $Rh(acac)(CO)_2$, $[Rh(COD)Cl]_2$ and $[Rh(NBD)Cl]_2$. SiCAC reactions afforded 1,3,6,8-tetrahydrobenzo[1,2-c:3,4-c']difurans or 1,2,3,6,7,8-hexahydropyrrolo[3,4-e]isoindoles as a mixture of two products: the 4-triorganosilyl-substituted compound and the corresponding desilylated product (Scheme 33) [88].

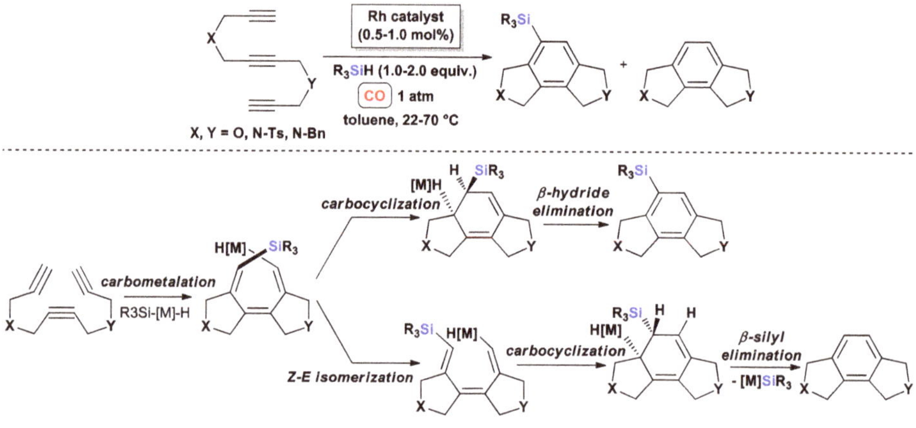

Scheme 33. Rh-catalyzed *cascade* SiCAC of oxygen- and nitrogen-containing analogs of dodec-1,6,11-triynes: synthesis of 1,3,6,8-tetrahydrobenzo[1,2-c:3,4-c']difurans and 1,2,3,6,7,8-hexahydropyrrolo[3,4-e]isoindoles.

The most plausible mechanism starts with a silicon-initiated cascade carbometalation to give a 3,3'-bifuranylidene/3,3'-bipyrrolylidene intermediate, which can then follow two different pathways: (a) carbocyclization followed by β-hydride elimination, affording the 4-triorganosilyl-substituted product; (b) a Z-E isomerization favored by high temperatures, followed by a similar carbocyclization step and subsequent β-silyl elimination, giving the corresponding desilylated product.

3.3. Synthesis of Heterocycles Via Metal-Catalyzed Heteroatom-Promoted Silylcarbocyclizations of Alkynes

As already stressed, carbocyclizations of alkynes are extremely important reactions in the synthesis of numerous carbonyl and heterocyclic compounds of pharmaceutical and theoretical interest.

During the studies on the mechanism and the synthetic potential of silylformylation reactions of acetylenes, new and interesting reactions of *heteroatom-promoted* silylcarbocyclization were discovered [89,90]. Reactions of propargyl alcohols or amides with a hydrosilane, catalyzed by $Rh_4(CO)_{12}$, and in the presence of a base (e.g., Et_3N, DBU) provided as main products (triorganosilyl)methylene- β-lactones and β-lactams respectively, which are important scaffolds present in many natural compounds.

3.3.1. Heteroatom-Promoted Silylcarbocyclizations of Ethynyl Alcohols

The first example of *heteroatom-promoted* silylcarbocyclizations of propargyl alcohols was described by Matsuda and co-workers in 1990 [89]. Based on a previous study on the silylformylation reactions of functionalized alkynes [16], they decided to investigate the possible cyclization of acetylenic alcohols under the silylformylation reactions conditions (R_3SiH, Et_3N, CO, 100 °C, $Rh_4(CO)_{12}$) (Scheme 34). Linear and branched alcohols were tested in the presence of different silanes (Me_2PhSiH, t-$BuMe_2SiH$, Et_3SiH, $(i$-$Pr)_3SiH$) and bases (Et_3N, DBU, pyridine, DABCO, DBU). Chemoselectivity of the reaction (i.e., β-lactone vs. aldehyde) depended strongly on the steric hindrance of silane and on the strength of the base. Indeed, while the reaction between 2-propynol (Scheme 34, R^1, R^2 = H) and Me_2PhSiH in the presence of Et_3N and $Rh_4(CO)_{12}$ gave exclusively the corresponding alkenal, the use of t-$BuMe_2SiH$ and DBU afforded the expected methylene-β-lactones in very high yields (79–86%) and selectivity.

The formation of two different products, the alkenal and β-lactones ring, was explained by Matsuda and co-workers with the hypothesis of a Rh-acyl species (Figure 1), which could be the common intermediate to give both products. Indeed, an experiment performed under carbonylation conditions of the alkenal did not afford the corresponding lactone derivatives, thus suggesting that the two products are formed competitively.

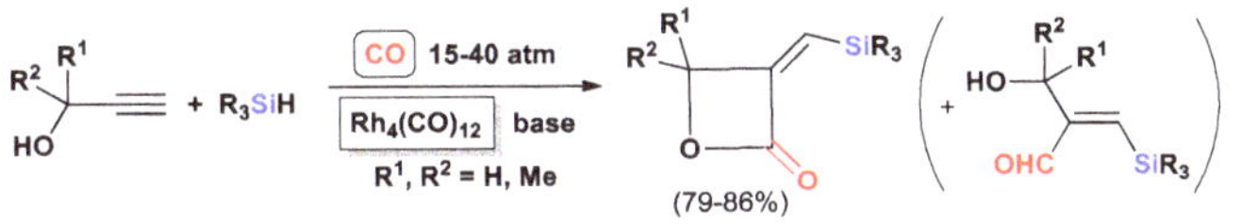

Scheme 34. First example of *heteroatom-promoted* silylcarbocyclization of ethynyl alcohols.

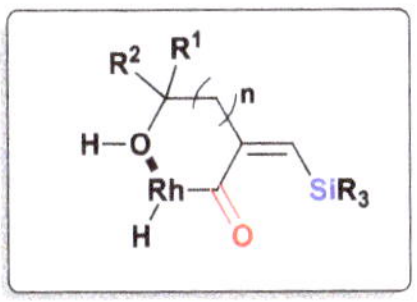

Figure 1. Rhodium-acyl intermediate hypothesized by Matsuda and co-workers in the *heteroatom-promoted* SiCAC of propargyl alcohols.

Heteroatom-promoted SiCAC was then applied to the synthesis of spiro-type β-lactones, which required DBU as the base (Scheme 35) to generate the desired compounds in good yields (68–86%) [89,91].

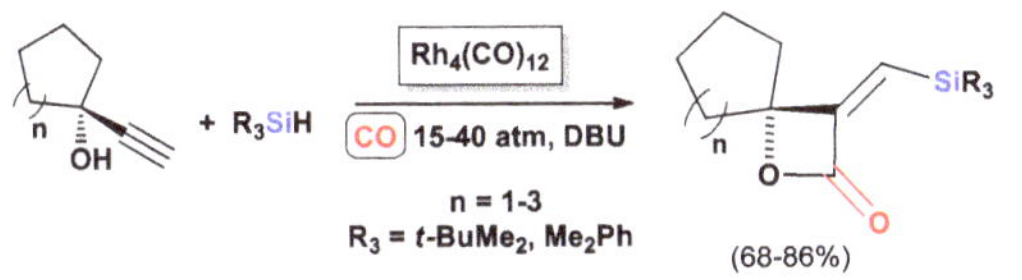

Scheme 35. Synthesis of spiro-type β-lactones via *heteroatom-promoted* SiCAC of ethynylcycloalkanols.

Moreover, the same SiCAC reaction was also extended to butynol and pentynols derivatives, affording the corresponding γ- and δ-lactones in very high yields (84–90%) even if Et$_3$N was used, thus indicating that the formation of both five- and six-membered heterocyclic compounds is extremely favored (Scheme 36) [89].

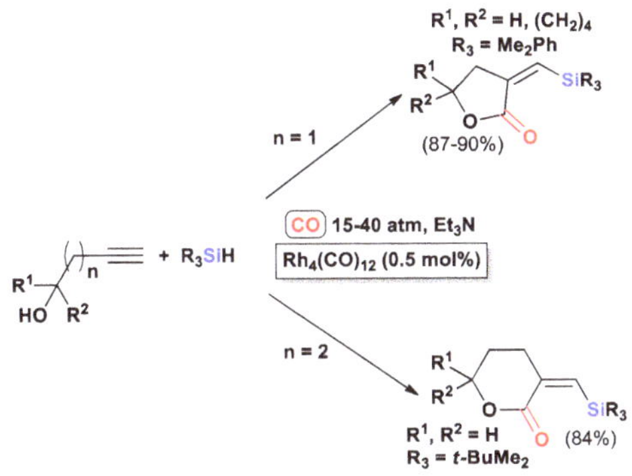

Scheme 36. Synthesis of γ- and δ-lactones via *heteroatom-promoted* SiCAC reactions.

Similarly, complete chemoselectivity towards the lactone formation was observed by the same research group in the silylcarbocyclization reactions of cyclohexanol containing a propynyl or butynyl group connected to alcohol carbon atom (Scheme 37) [4]. Once more, the use of DBU together with *t*-BuMe$_2$SiH and Rh$_4$(CO)$_{12}$ yielded the corresponding γ- and δ-spirolactones, selectively.

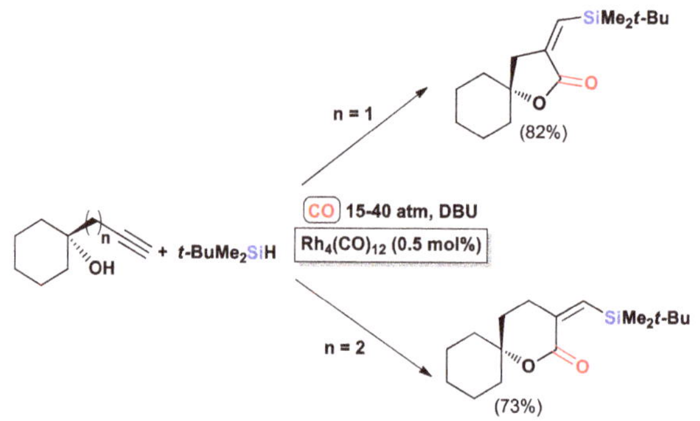

Scheme 37. Synthesis of γ- and δ-spirolactones via *heteroatom-promoted* SiCAC reactions.

However, γ-lactones can also be obtained via SiCAC reaction of *trans*-2-ethynylcyclopentanol and cyclohexanol, as depicted in Scheme 38 [4]. DBU and *t*-BuMe$_2$SiH, together with Rh$_4$(CO)$_{12}$ as catalyst, resulted in being the best reagents for the selective formation of the lactone nucleus fused to a cyclopentane and cyclohexane ring in high yields (83–84%).

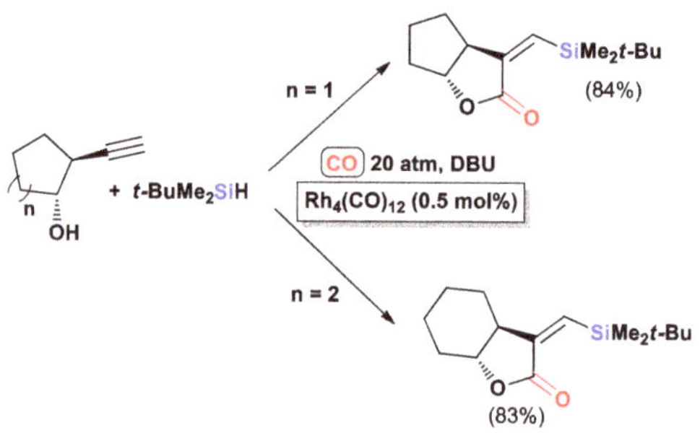

Scheme 38. *Heteroatom-promoted* SiCAC reaction applied to the synthesis of γ-lactones.

A few years later, Aronica and co-workers reported a detailed study on the *heteroatom-promoted* SiCAC reaction of several propargyl alcohols characterized by different steric and electronic requirements [92]. All reactions were performed in CH$_2$Cl$_2$, at 100 °C, under 30 atm of CO, with 0.1 mol% of Rh$_4$(CO)$_{12}$ as catalyst and DBU as base. As previously observed by Matsuda, the chemoselectivity of the process was clearly influenced by steric hindrance: the presence of a *tert*-butyl, ethyl or cyclohexyl group on the propargyl carbon atom determined a nearly total chemoselectivity towards β-lactones, while the reaction involving 3-propynol generated the corresponding β-silylalkenal predominantly (Scheme 39).

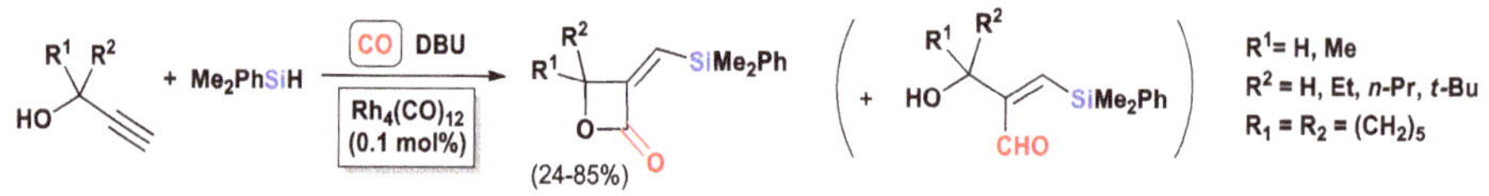

Scheme 39. *Heteroatom-promoted* SiCAC of propargyl alcohols.

In order to improve the formation of the lactone ring, arylsilanes containing a hindered substituent (MePh$_2$SiH, Ph$_3$SiH, *o*-CH$_3$-(C$_6$H$_4$)Me$_2$SiH, *p*-Ph-(C$_6$H$_4$)Me$_2$SiH) were tested in the *heteroatom-promoted* silylcarbocyclization of 1-hexynol performed with Rh$_4$(CO)$_{12}$ as catalyst (Scheme 40) [92].

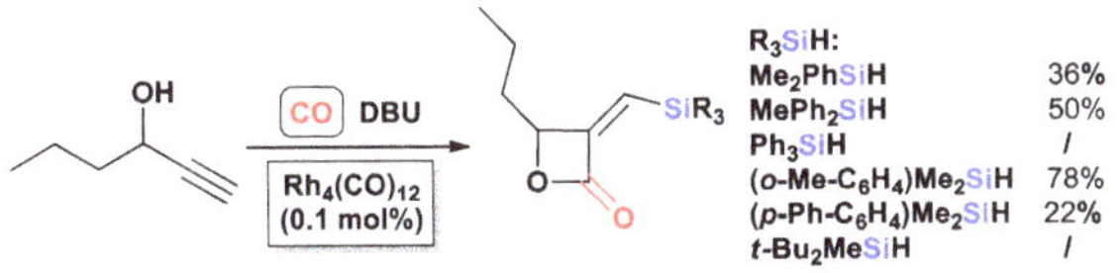

Scheme 40. *Heteroatom-promoted* SiCAC protocol applied to 1-hexyn-3-ol involving arylsilanes with a hindered substituent.

The obtained results clearly indicated that the choice of hydrosilane plays a crucial role. Indeed, the best chemo-selectivity was observed in the reaction with *o*-tolyldimethylsilane and diphenylmethylsilane; on the contrary, Ph$_3$SiH and (*t*-Bu)$_2$PhSiH were totally inactive. Finally moving from dichloromethane to toluene as solvent and operating at lower temperature (70 °C), a significant improvement of the lactone selectivity was detected. The same results were obtained when two homopropargyl alcohols were considered. In these cases, the cyclization process was definitely favored (Scheme 41). All (dimethylphenylsilyl)methylene β- and γ-lactones can be submitted to a TBAF-promoted phenyl migration without a ring opening, affording useful building blocks for the synthesis of pharmaceutical compounds.

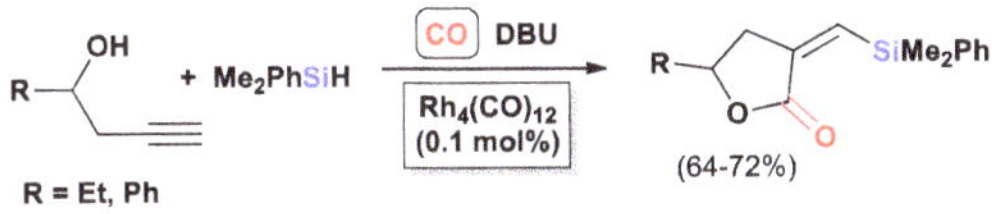

Scheme 41. *Heteroatom-promoted* SiCAC of homopropargyl alcohols: synthesis of γ-lactones.

The same authors then investigated the *heteroatom-promoted* silylcarbocyclization of propargyl alcohols promoted by different catalytic species [93]. Initially, a preliminary study on Rh/mesitylene co-condensate, prepared according to the MVS technique [35,51–53] and consisting of small rhodium nanoclusters, was carried out. With respect to commercial Rh$_4$(CO)$_{12}$, Rh/mesitylene catalyst showed excellent performance in the SiCAC process of 1-hexyn-3-ol with *t*-BuMe$_2$SiH, in CH$_2$Cl$_2$ and DBU, at 100 °C and under 30 atm of carbon monoxide. The β-lactone ring was obtained with 87% of selectivity (Scheme 42).

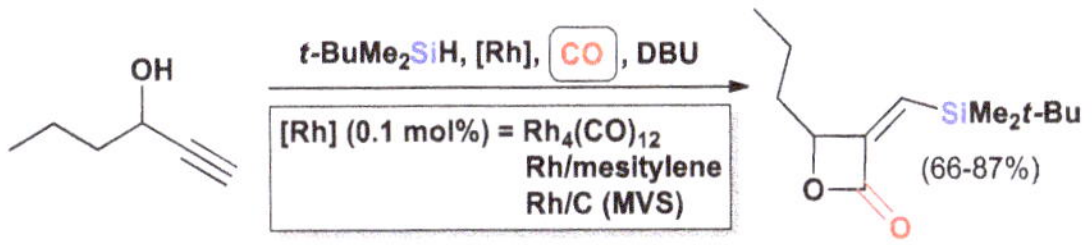

Scheme 42. *Heteroatom-promoted* SiCAC reaction of 1-hexyn-3-ol promoted by different catalytic species.

When the same rhodium co-condensate was deposited on several matrices (charcoal, γ-alumina, Fe$_2$O$_3$ and polybenzoimidazole), supported Rh/C, Rh/γ-Al$_2$O$_3$, Rh/Fe$_2$O$_3$ and Rh/PBI were prepared and tested in the *heteroatom-promoted* silylcarbocyclization reactions [93]. Among them, Rh/C showed the best results in terms of conversion (87%) and selectivity (92%), even compared with a commercial Rh/C species. As a consequence, Rh/C (MVS) was used in the SiCAC processes of 3-dialkylpropargyl alcohols with Me$_2$PhSiH, in CH$_2$Cl$_2$ as solvent and DBU as base (Scheme 43): the reactions afforded β-lactone derivatives with almost complete chemoselectivity (92–97%). High resolution transmission electron microscopy (HR-TEM) analysis of Rh/C (MVS) indicated the presence of very small Rh nanoparticles (2.4 nm mean diameter) on the support, which could be the reason of its high catalytic activity. Unfortunately, preliminary investigations evidenced a relevant metal leaching into solution during the reactions, thus indicating that Rh/C (MVS) acted as a reservoir of soluble active nanoparticles.

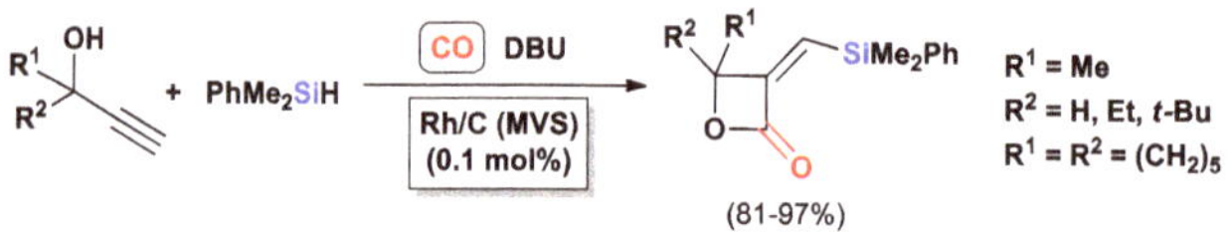

Scheme 43. *Heteroatom-promoted* SiCAC of 3-dialkylpropargyl alcohols catalyzed by Rh/C (MVS).

In 2017, the Aronica's research group developed a new protocol for the synthesis of 3-isochromanone derivatives based on *heteroatom-promoted* silylcarbocyclization reactions of 2-ethynylbenzyl alcohol [94]. Initially, the SiCAC process was performed with Me$_2$PhSiH as hydrosilane and Rh$_4$(CO)$_{12}$ as catalyst, in CH$_2$Cl$_2$ as solvent and DBU as base, under 30 atm of CO at 100 °C. Surprisingly, together with the expected product, relevant amounts of the corresponding hydrogenated by-product were obtained, regardless of catalyst loading, temperature and CO pressure (Scheme 44), probably due to the formation of hydrogen during the SiCAC reaction [89]. Only a slight improvement in chemoselectivity (33% isochromanone) was observed when $(\eta^6\text{-}C_6H_6BPh_3)^-Rh^+(1,5\text{-}COD)$ (Rh^{sw}) catalyst was employed instead of Rh$_4$(CO)$_{12}$.

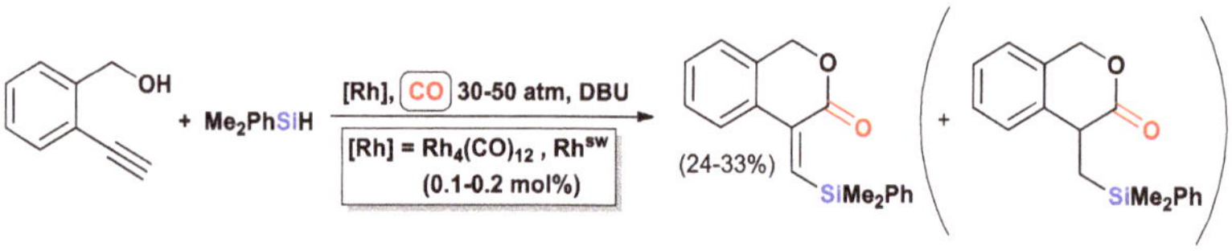

Scheme 44. *Heteroatom-promoted* SiCAC of 2-ethynylbenzyl alcohol with Me$_2$PhSiH for the synthesis of 3-isochromanone derivatives.

Unexpectedly, working without DBU, the selectivity towards methyleneisochromanone increased. As a consequence, the optimized experimental conditions (Rh^{sw} 0.1–0.2 mol%, 100 °C, 30–50 atm of CO, 2–6 h), were used for the SiCAC reactions of ethynylbenzyl alcohol with different aryldimethylsilanes. All reactions afforded the expected products with good yields and total stereoselectivity, since only (Z) isochromanones were formed (Scheme 45).

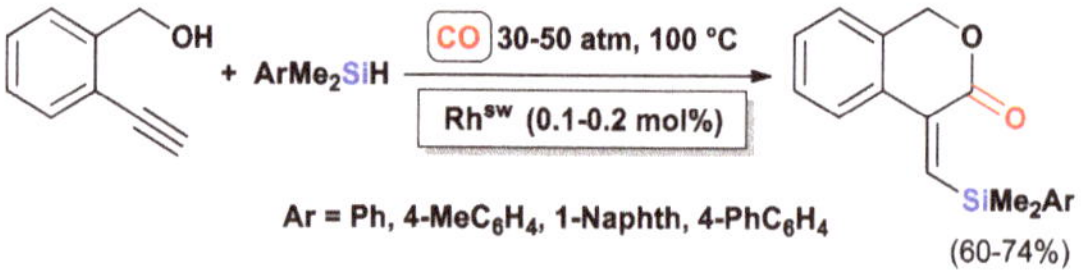

Scheme 45. *Heteroatom-promoted* SiCAC of 2-ethynylbenzyl alcohol with different aryldimethylsilanes for the synthesis of (Z)-isochromanones.

3.3.2. Heteroatom-Promoted Silylcarbocyclizations of Ethynyl Amines

The β-lactam moiety is the key of one of the most widely employed class of antibiotics [95–97], i.e., β-lactam antibiotics such as penicillins and cephalosporins, which are distinguished by good tolerance and therapeutic safety. In particular, the α-methylene-β-lactam unit is a very common structural feature included in potent β-lactamase inhibitors [98], such as asparenomycins [99,100] and penicillanic acids [101,102]. Therefore, the synthesis of α-methylene-β-lactams (3-methylene-2-azetidinones) has received great attention in the literature [103–106].

In 1991, Matsuda et al. described the first example of *heteroatom-promoted* SiCAC reactions of ethynyl amines, applied to the formation of α-silylmethylene-β-lactams [90]. On the base of the results previously obtained in the silylcarbocyclizations of propargyl alcohols, they started their investigation with the reaction of N-(1-ethynylcyclohexyl)-p-toluensulfonamide with RMe$_2$SiH (R = Ph or t-Bu), Rh$_4$(CO)$_{12}$, a suitable base, at 100 °C and under 20 atm of CO. In particular, the best result (81% lactam) was obtained by the combined use of a bulky silane (*t*-BuMe₂SiH) and DBU as the base (Scheme 46). Under the same experimental conditions, other sulfonamides afforded the corresponding β-lactams with good selectivity; instead, the less hindered toluensulfonamide and *N*-propargylcarbamates generated the corresponding silylformylation product predominantly.

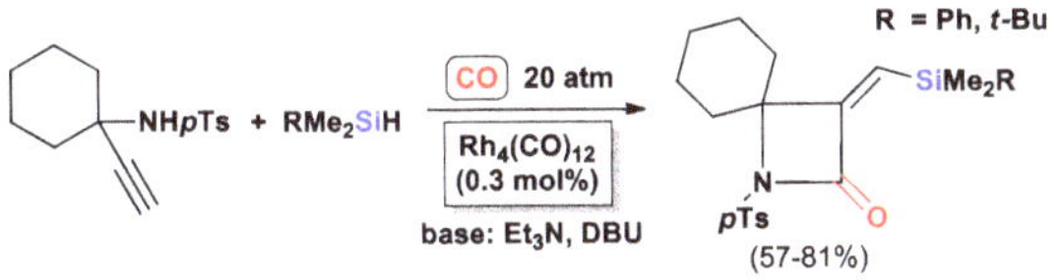

Scheme 46. First example of *heteroatom-promoted* silylcarbocyclization of propargyl amides: synthesis of α-silylmethylene-β-lactams.

Better results were described by the same research group when alkynylbenzylamines were used as substrates for *heteroatom-promoted* silylcarbocyclizations, which generated the corresponding γ- and δ-lactams [4]. The reactions were carried out with Rh₄(CO)₁₂ (0.25 mol%), DBU, 20 atm of CO, 100 °C and *t*-BuMe₂SiH, which was fundamental for the cyclization reaction to occur (Scheme 47).

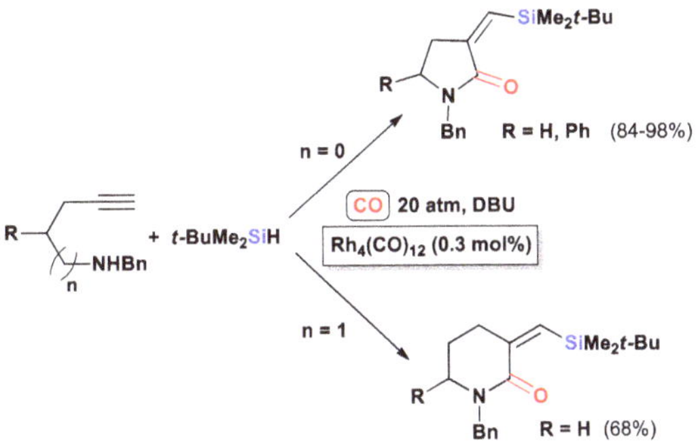

Scheme 47. *Heteroatom-promoted* SiCAC of benzylamines applied to the synthesis of γ- and δ-lactams.

Ethynylpiperidine derivatives were also tested in the *heteroatom-promoted* SiCAC reaction under the same experimental conditions, affording the corresponding ring-fused lactam compounds in good yields (Scheme 48) [4].

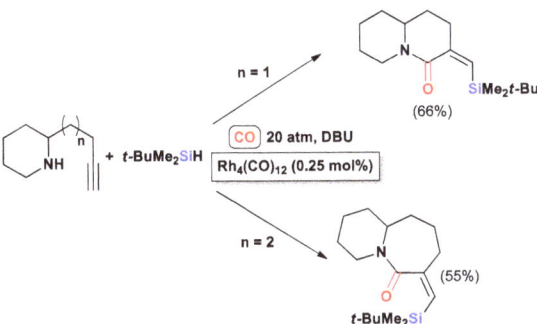

Scheme 48. *Heteroatom-promoted* SiCAC reactions of ethynylpiperidines.

Isoquinoline-based substrates were deeply investigated in *heteroatom-promoted* silylcarbocyclizations. Matsuda et al. worked under the above-mentioned optimized conditions (i.e., *t*-BuMe₂SiH, Rh₄(CO)₁₂

0.25 mol%, DBU, CO 20 atm, 100 °C), affording selectively the corresponding benzoindolizidinone as the Z-isomer (Scheme 49, path A) [4].

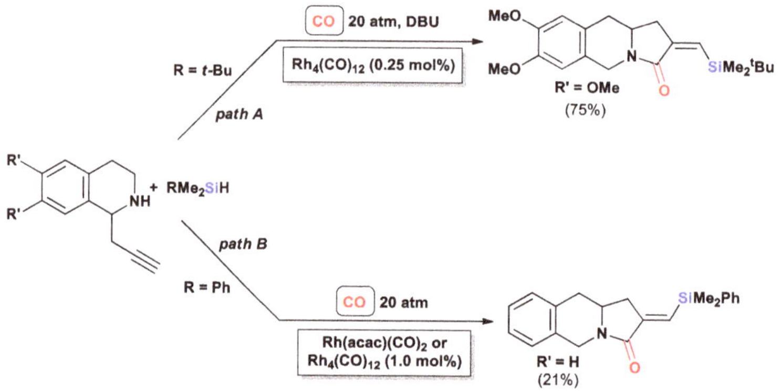

Scheme 49. *Heteroatom-promoted* SiCAC reactions applied to the synthesis of benzoindolizidinones.

Ojima and co-workers used different conditions: PhMe$_2$SiH, Rh(acac)(CO)$_2$ 1 mol%, in toluene under 50 atm of carbon monoxide, at 60 °C but without DBU. In this case, SiCAC took place with different stereoselectivity, giving the *E*-isomer of benzoindolizidinone in poor yield (21%), probably due to the absence of the base (Scheme 49, path B) [107]. The authors suggested that the isomerization of silylvinyl group took place during the reaction (Scheme 50), after the addition of H[Rh]SiMe$_2$Ph to the triple bond, according to what was previously observed in the hydrosilylation reaction of 1-alkynes [108].

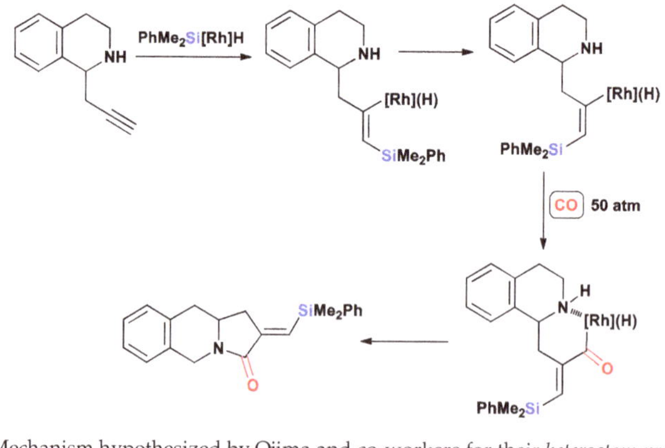

Scheme 50. Mechanism hypothesized by Ojima and co-workers for their *heteroatom-promoted* SiCAC protocol of isoquinoline-based substrates.

Prompted from this result, the same group investigated the formation of an indolizidine skeleton by means of *heteroatom-promoted* SiCAC of butynylpyrrolidine and hydrosilanes [107]. Both Rh(acac)(CO)$_2$ and Rh$_2$Co$_2$(CO)$_{12}$ (2 mol%) were an effective catalyst for the silylcarbocyclization reaction, which was found to be very sensitive to the nature of the silane (Scheme 51).

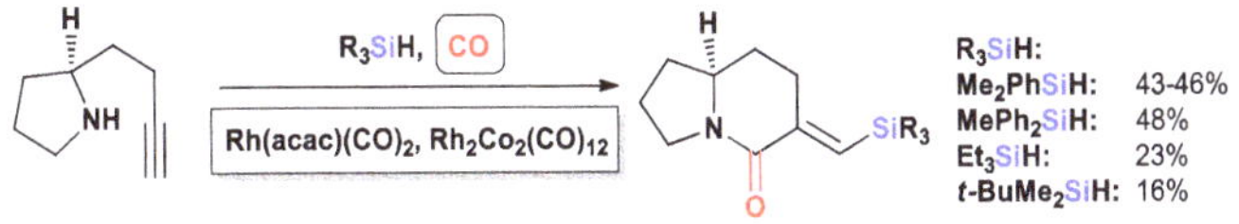

Scheme 51. *Heteroatom-promoted* SiCAC reactions applied to the synthesis of indolizidinones.

On the contrary, when Matsuda et al. investigated the reactivity of 5-(prop-2-yn-1-yl)pyrrolidin-2-one in silylcarbocyclization reaction, the corresponding Z-pyrrolizine-3,5(2H,6H)-dione was exclusively found, thus confirming again the central role of the base (Scheme 52).

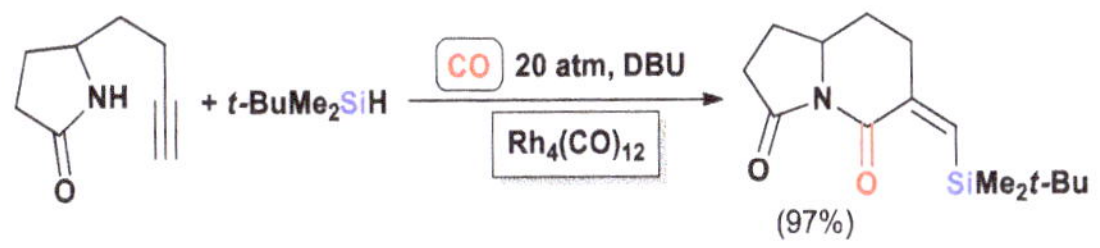

Scheme 52. *Heteroatom-promoted* SiCAC applied to the synthesis of pyrrolizine-3,5(2H,6H)-dione.

More recently, Aronica et al. applied the *heteroatom-promoted* SiCAC transformation to some propargyl tosylamides, using $Rh_4(CO)_{12}$ (0.1 mol%) as catalyst, DBU as base, under CO pressure (30 atm), at 100 °C [36]. The presence of DBU and a quaternary α-carbon on the substrate were essential for the SiCAC reaction to occur with complete chemoselectivity towards the β-lactam ring, regardless the steric and electronic requirements of the silanes. Moreover, (Z)-stereoisomers were exclusively obtained (Scheme 53).

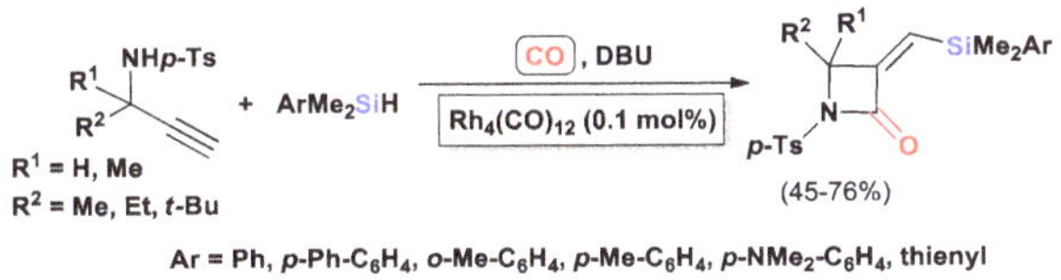

Scheme 53. *Heteroatom-promoted* SiCAC of propargyl tosylamides for the synthesis of β-lactams.

Taking into account the same tosylamides, Aronica and co-workers tested different supported Rh catalysts prepared according to the MVS technique in the SiCAC reactions [93]. As already observed for the reactions performed with propargyl alcohols, among MVS Rh/C, Rh/γ-Al_2O_3, Rh/ Fe_2O_3 and Rh/PBI, the first species showed a specific activity even higher than homogeneous $Rh_4(CO)_{12}$ used as a reference catalyst. When Rh/C was used in the *heteroatom-promoted* SiCAC, the expected β-lactams were achieved with 90–100% selectivity after 4 h, at 100 °C and 30 atm CO (Scheme 54). Moreover, the same batch of Rh/C could be reused without loss of activity.

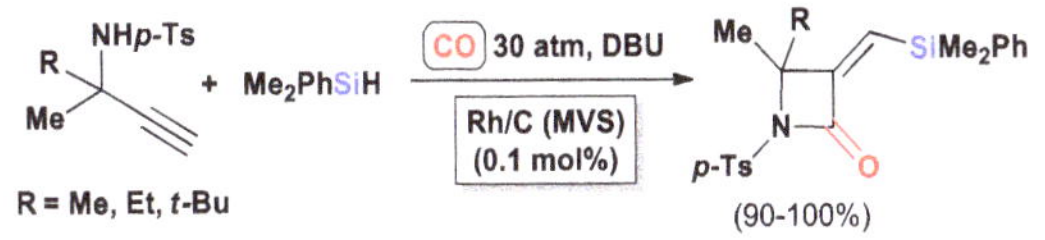

Scheme 54. Synthesis of β-lactams through *heteroatom-promoted* SiCAC reactions of propargyl tosylamides catalyzed by Rh/C (MVS).

In 2019, Albano and co-workers described the first synthesis of indolines and tetrahydroisoquinolines via *heteroatom-promoted* SiCAC of suitable tosylamides [109]. The (2-ethynylphenyl)-4-tosylamide was first tested in the reaction with dimethylphenylsilane, promoted by Rh^{SW} (0.3 mol%), under 30 atm of CO, at 30 °C (Scheme 55). The formation of the corresponding five-membered heterocyclic compound took place without the need of a base as previously observed in the silylcarbocyclization applied to the synthesis of isochromanones [94]. Together with tosylindolinone, the corresponding hydrogenated derivative tosylindolinol was surprisingly generated too. In order to improve the chemoselectivity of the SiCAC process, reactions at higher temperature (50–100 °C) were performed. However, tosylindolynol was obtained as the sole product.

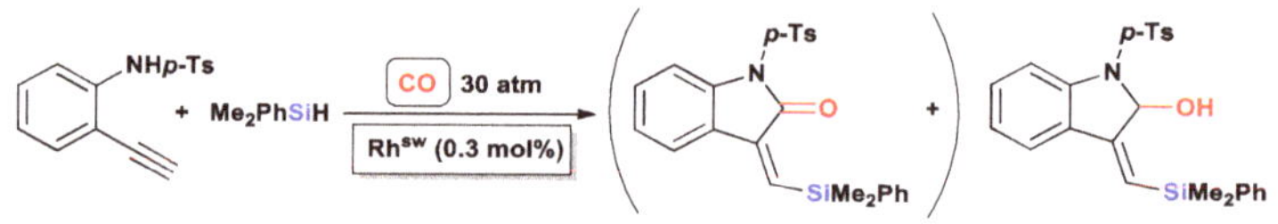

Scheme 55. Synthesis of tosylindolinol via *heteroatom-promoted* SiCAC of (2-ethynylphenyl)-4-tosylamide.

The silylcarbocyclization was then extended to the synthesis of tetrahydroisoquinolines but, again, tosyltetrahydroisoquinolinols were selectively obtained regardless the temperature employed and the nature of the catalyst (i.e., Rh^{sw}, Rh(acac)(CO)$_2$, Rh$_6$(CO)$_{16}$), as depicted in Scheme 56 [109]. The formation of the reduced compounds was ascribed to the presence of molecular H$_2$, which is formed as a by-product in the reaction vessel.

Scheme 56. *Heteroatom-promoted* SiCAC applied to the synthesis of tosyltetrahydroisoquinolinols.

With the optimal reaction conditions in hand, the authors investigated the reactivity of hydrosilanes possessing different steric requirements with both tosylamides. The optimized SiCAC procedure afforded the corresponding products in very good yields (51–75%) [109]. Moreover, the ring formation took place with complete stereoselectivity of the exocyclic double bond (Z isomer exclusively), not only in reactions involving aryl silanes but also for benzyl derivative (Scheme 57). The obtained silylated tosylindolinols and tosyltetrahydroisoquinolinols could be easily desilylated by means of TBAF, which promoted aryl rearrangements from silicon to the adjacent carbon atom, generating new polyfunctionalized N-heterocycles.

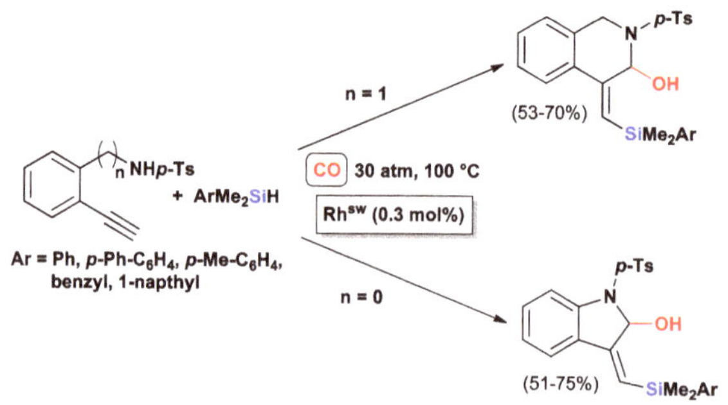

Scheme 57. Synthesis of tosylindolinols and tosyltetrahydroisoquinolinols via *heteroatom-promoted* SiCAC with several ArMe$_2$SiH.

4. Conclusions

In summary, intramolecular silylformylation and silylcarbocyclization processes provide efficient and versatile methods for the construction of monocyclic, bicyclic and polycyclic heterocycles.

We really hope that the present review may stimulate further research in the field of silylformylation and silylcarbocyclization reactions, and in particular for the preparation of new biologically relevant O-

and N-heterocycles, as a valid alternative to the most common cycloaddition [110–112] or cross-coupling reactions [113–116]. Moreover, the recent studies on CO surrogates [117–119] may be a strong stimulus for innovative and safer development of new silylcarbonylation processes.

Author Contributions: Manuscript outline, G.A. and L.A.A.; bibliographic material selection, G.A. and L.A.A.; manuscript writing—original draft preparation, review and editing, G.A. and L.A.A. All authors have read and agreed to the published version of the manuscript.

Funding: This research received no external funding.

Conflicts of Interest: The authors declare no conflict of interest.

References

1. Chatani, N.; Murai, S. HSiR$_3$/CO as the potent reactant combination in developing new transition-metal-catalyzed reactions. *Synlett* **1996**, *1996*, 414–424. [CrossRef]
2. Matsuda, I.; Fukuta, Y.; Tsuchihashi, T.; Nagashima, H.; Itoh, K. Rhodium-catalyzed silylformylation of Acetylenic bonds: Its scope and mechanistic considerations. *Organometallics* **1997**, *16*, 4327–4345. [CrossRef]
3. Leighton, J.L. Stereoselective rhodium(I)-catalyzed hydroformylation and silylformylation reactions and their application to organic synthesis. In *Modern Rhodium-Catalyzed Organic Reactions*; Evans, P.A., Ed.; WILEY-VCH Verlag GmbH & Co. KGaA: Weinheim, Germany, 2005; pp. 93–110.
4. Matsuda, I. Silylformylation. In *Comprehensive Organometallic Chemistry III*; Mingos, D.M.P., Crabtree, R.H., Eds.; Elsevier: Amsterdam, The Netherlands, 2007; pp. 473–510.
5. Aronica, L.A.; Caporusso, A.M.; Salvadori, P. Synthesis and reactivity of silylformylation products derived from alkynes. *Eur. J. Org. Chem.* **2008**, *2008*, 3039–3060. [CrossRef]
6. Beller, M.; Cornils, B.; Frohning, C.D.; Kohlpaintner, C.W. Progress in hydroformylation and carbonylation. *J. Mol. Catal. A Chem.* **1995**, *104*, 17–85. [CrossRef]
7. Reiser, O. Metal-catalyzed hydroformylations. In *Organic Synthesis Highlights IV*; Schmalz, H.-G., Ed.; WILEY-VCH Verlag GmbH: Weinheim, Germany, 2000; pp. 97–103.
8. Ojima, I.; Tsai, C.-Y.; Tzamarioudaki, M.; Bonafoux, D. The hydroformylation reaction. In *Organic Reactions*; Overman, L.E., Ed.; John Wiley & Sons, Inc.: Hoboken, NJ, USA, 2000; Volume 56.
9. Breit, B.; Seiche, W. Recent advances on chemo-, regio-and stereoselective hydroformylation. *Synthesis* **2001**, *2001*, 0001–0036. [CrossRef]
10. Wiese, K.-D.; Obst, D. Hydroformylation. *Top. Organomet. Chem.* **2006**, *18*, 1–33.
11. Franke, R.; Selent, D.; Börner, A. Applied hydroformylation. *Chem. Rev.* **2012**, *112*, 5675–5732. [CrossRef]
12. Pospech, J.; Fleischer, I.; Franke, R.; Buchholz, S.; Beller, M. Alternative metals for homogeneous catalyzed hydroformylation reactions. *Angew. Chem. Int. Ed.* **2013**, *52*, 2852–2872. [CrossRef]
13. Breit, B.; Diab, L. Hydroformylation and related carbonylation reactions of alkenes, alkynes, and allenes. In *Comprehensive Organic Synthesis*, 2nd ed.; Knochel, P., Molander, G.A., Eds.; Elsevier: Amsterdam, The Netherlands, 2014; Volume 4, pp. 995–1053.
14. Wu, X.-F.; Fang, X.; Wu, L.; Jackstell, R.; Neumann, H.; Beller, M. Transition-metal-catalyzed carbonylation reactions of olefins and alkynes: A personal account. *Acc. Chem. Res.* **2014**, *47*, 1041–1053. [CrossRef]
15. Hanf, S.; Alvarado Rupflin, L.; Gläser, R.; Schunk, S.A. Current state of the art of the solid rh-based catalyzed hydroformylation of short-chain olefins. *Catalysts* **2020**, *10*, 510. [CrossRef]
16. Matsuda, I.; Ogiso, A.; Sato, S.; Izumi, Y. An efficient silylformylation of alkynes catalyzed by tetrarhodium dodecacarbonyl. *J. Am. Chem. Soc.* **1989**, *111*, 2332–2333. [CrossRef]
17. Aronica, L.A.; Raffa, P.; Valentini, G.; Caporusso, A.M.; Salvadori, P. Silylformylation—Fluoride-assisted aryl migration of acetylenic derivatives in a versatile approach to the synthesis of polyfunctionalised compounds. *Eur. J. Org. Chem.* **2006**, *2006*, 1845–1851. [CrossRef]
18. Aronica, L.A.; Raffa, P.; Caporusso, A.M.; Salvadori, P. Fluoride-promoted rearrangement of organo silicon compounds: A new synthesis of 2-(arylmethyl)aldehydes from 1-alkynes. *J. Org. Chem.* **2003**, *68*, 9292–9298. [CrossRef] [PubMed]
19. Aronica, L.A.; Raffa, P.; Valentini, G.; Caporusso, A.M.; Salvadori, P. Silylformylation–desilylation of propargyl amides: Synthesis of α,β-unsaturated aldehydes. *Tetrahedron Lett.* **2006**, *47*, 527–530. [CrossRef]

20. Donskaya, N.A.; Yur'eva, N.M.; Sigeev, A.S.; Voevodskaya, T.I.; Beletskaya, I.P.; Tretyakov, V.F. Silylformylation of alkynes catalysed by Di-μ-chlorotetrakis(η2-methylene-cyclopropane)dirhodium. *Mendeleev Commun.* **1995**, *5*, 220–221. [CrossRef]
21. Alonso, M.A.; Casares, J.A.; Espinet, P.; Vallés, E.; Soulantica, K. Efficient silylformylation of alkynes catalyzed by rhodium complexes with P,N donor ligands. *Tetrahedron Lett.* **2001**, *42*, 5697–5700. [CrossRef]
22. Basato, M.; Biffis, A.; Martinati, G.; Zecca, M.; Ganis, P.; Benetollo, F.; Aronica, L.A.; Caporusso, A.M. Cationic carboxylato complexes of dirhodium(II) with oxo thioethers: Promising catalysts with unusual coordination modes. *Organometallics* **2004**, *23*, 1947–1952. [CrossRef]
23. Basato, M.; Biffis, A.; Martinati, G.; Tubaro, C.; Graiff, C.; Tiripicchio, A.; Aronica, L.A.; Caporusso, A.M. Cationic complexes of dirhodium(II) with 1,8-naphthyridine: Catalysis of reactions involving silanes. *J. Organomet. Chem.* **2006**, *691*, 3464–3471. [CrossRef]
24. Biffis, A.; Conte, L.; Tubaro, C.; Basato, M.; Aronica, L.A.; Cuzzola, A.; Caporusso, A.M. Highly selective silylformylation of internal and functionalised alkynes with a cationic dirhodium(II) complex catalyst. *J. Organomet. Chem.* **2010**, *695*, 792–798. [CrossRef]
25. Ojima, I.; Ingallina, P.; Donovan, R.J.; Clos, N. Silylformylation of 1-hexyne catalyzed by rhodium-cobalt mixed-metal carbonyl clusters. *Organometallics* **1991**, *10*, 38–41. [CrossRef]
26. Ojima, I.; Donovan, R.J.; Eguchi, M.; Shay, W.R.; Ingallina, P.; Korda, A.; Zeng, Q. Silylformylation catalyzed by Rh and Rh-Co mixed metal complexes and its application to the synthesis of pyrrolizidine alkaloids. *Tetrahedron* **1993**, *49*, 5431–5444. [CrossRef]
27. Ojima, I.; Donovan, R.J.; Ingallina, P.; Clos, N.; Shay, W.R.; Eguchi, M.; Zeng, Q.; Korda, A. Organometallic chemistry and homogeneous catalysis of Rh and Rh-Co mixed metal carbonyl clusters. *J. Cluster Sci.* **1992**, *3*, 423–438. [CrossRef]
28. Ojima, I.; Zhaoyang, L.; Donovan, R.J.; Ingallina, P. On the mechanism of silylformylation catalyzed by Rh-Co mixed metal complexes. *Inorg. Chim. Acta* **1998**, *270*, 279–284. [CrossRef]
29. Yoshikai, N.; Yamanaka, M.; Ojima, I.; Morokuma, K.; Nakamura, E. Bimetallic synergism in alkyne silylformylation catalyzed by a cobalt–rhodium mixed-metal cluster. *Organometallics* **2006**, *25*, 3867–3875. [CrossRef]
30. Doyle, M.P.; Shanklin, M.S. Highly Regioselective and stereoselective silylformylation of alkynes under mild conditions promoted by dirhodium(II) perfluorobutyrate. *Organometallics* **1994**, *13*, 1081–1088. [CrossRef]
31. Doyle, M.P.; Shanklin, M.S. Highly efficient regioselective silylcarbonylation of alkynes catalyzed by dirhodium(II) perfluorobutyrate. *Organometallics* **1993**, *12*, 11–12. [CrossRef]
32. Zhou, Z.; Facey, G.; James, B.R.; Alper, H. Interconversion between zwitterionic and cationic rhodium(I) complexes of demonstrated value as catalysts in hydroformylation, silylformylation, and hydrogenation reactions. dynamic 31P{1H} NMR studies of (η6-PhBPh3)-Rh+(DPPB) and [Rh(DPPB)2]+BPh4-in solution. *Organometallics* **1996**, *15*, 2496–2503.
33. Zhou, J.-Q.; Alper, H. Zwitterionic rhodium(I) complex catalyzed net silylhydroformylation of terminal alkynes. *Organometallics* **1994**, *13*, 1586–1591. [CrossRef]
34. Okazaki, H.; Kawanami, Y.; Yamamoto, K. The Silylformylation of simple 1-alkynes catalyzed by [Rh(cod)][BPh4] in an Ionic liquid, [Bmim][PF6], under biphasic conditions: An efficiently reusable Catalyst system. *Chem. Lett.* **2001**, *30*, 650–651. [CrossRef]
35. Aronica, L.A.; Terreni, S.; Caporusso, A.M.; Salvadori, P. Silylformylation of chiral 1-alkynes, catalysed by solvated rhodium atoms. *Eur. J. Org. Chem.* **2001**, *2001*, 4321–4329. [CrossRef]
36. Aronica, L.A.; Valentini, G.; Caporusso, A.M.; Salvadori, P. Silylation–desilylation of propargyl amides: Rapid synthesis of functionalised aldehydes and β-lactams. *Tetrahedron* **2007**, *63*, 6843–6854. [CrossRef]
37. Ojima, I.; Tzamarioudaki, M.; Tsai, C.-Y. Extremely chemoselective silylformylation and silylcarbocyclization of alkynals. *J. Am. Chem. Soc.* **1994**, *116*, 3643–3644. [CrossRef]
38. Carter, M.J.; Fleming, I. Allyl silanes in organic synthesis: Some reactions of 3-trimethylsilylcyclohex-4-ene-1,2-dicarboxylic acid and its derivatives. *J. Chem. Soc. Chem. Commun.* **1976**, 679–680. [CrossRef]
39. Fleming, I.; Perry, D.A. The synthesis of αβ-unsaturated ketones from β-silylenones and β-silylynones. *Tetrahedron* **1981**, *37*, 4027–4034. [CrossRef]
40. Jain, N.F.; Cirillo, P.F.; Schaus, J.V.; Panek, J.S. An efficient procedure for the preparation of chiral β-substituted (E)-crotylsilanes: Application of a rhodium(II) catalyzed silylformylation of terminal alkynes. *Tetrahedron Lett.* **1995**, *36*, 8723–8726. [CrossRef]

41. Eilbracht, P.; Hollmann, C.; Schmidt, A.M.; Bärfacker, L. Tandem silylformylation/Wittig olefination of terminal alkynes: Stereoselective synthesis of 2,4-dienoic esters. *Eur. J. Org. Chem.* **2000**, *2000*, 1131–1135. [CrossRef]
42. Bärfacker, L.; Hollmann, C.; Eilbracht, P. Rhodium(I)-catalysed hydrocarbonylation and silylcarbonylation reactions of alkynes in the presence of primary amines leading to 2-pyrrolidinones and 4-silylated 1-aza-1,3-butadienes. *Tetrahedron* **1998**, *54*, 4493–4506. [CrossRef]
43. Jung, M.E.; Gaede, B. Synthesis and diels-alder reactions of e-1-trimethylsilylbuta-1,3-diene. *Tetrahedron* **1979**, *35*, 621–625. [CrossRef]
44. Jones, T.K.; Denmark, S.E. Silicon-directed nazarov reactions II. Preparation and cyclization of β-silyl-substituted divinyl ketones. *Helv. Chim. Acta* **1983**, *66*, 2377–2396. [CrossRef]
45. Denmark, S.E.; Jones, T.K. Silicon-directed Nazarov cyclization. *J. Am. Chem. Soc.* **1982**, *104*, 2642–2645. [CrossRef]
46. Aronica, L.A.; Morini, F.; Caporusso, A.M.; Salvadori, P. New synthesis of α-benzylaldehydes from 2-(dimethylphenylsilylmethylene)alkanals by fluoride promoted phenyl migration. *Tetrahedron Lett.* **2002**, *43*, 5813–5815. [CrossRef]
47. Monteil, F.; Matsuda, I.; Alper, H. Rhodium-catalyzed intramolecular silylformylation of acetylenes: A vehicle for complete regio-and stereoselectivity in the formylation of acetylenic bonds. *J. Am. Chem. Soc.* **1995**, *117*, 4419–4420. [CrossRef]
48. Baldwin, J.E. Rules for ring closure. *J. Chem. Soc. Chem. Commun.* **1976**, 734–736. [CrossRef]
49. Gilmore, K.; Alabugin, I.V. Cyclizations of alkynes: Revisiting baldwin's rules for ring closure. *Chem. Rev.* **2011**, *111*, 6513–6556. [CrossRef] [PubMed]
50. Aronica, L.A.; Caporusso, A.M.; Salvadori, P.; Alper, H. Diastereoselective intramolecular silylformylation of ω-silylacetylenes. *J. Org. Chem.* **1999**, *64*, 9711–9714. [CrossRef]
51. Aronica, L.A.; Caporusso, A.M.; Tuci, G.; Evangelisti, C.; Manzoli, M.; Botavina, M.; Martra, G. Palladium nanoparticles supported on Smopex® metal scavengers as catalyst for carbonylative Sonogashira reactions: Synthesis of α,β-alkynyl ketones. *Appl. Catal. A Gen.* **2014**, *480*, 1–9. [CrossRef]
52. Albano, G.; Evangelisti, C.; Aronica, L.A. Hydrogenolysis of benzyl protected phenols and aniline promoted by supported palladium nanoparticles. *ChemistrySelect* **2017**, *2*, 384–388. [CrossRef]
53. Albano, G.; Interlandi, S.; Evangelisti, C.; Aronica, L.A. Polyvinylpyridine-supported palladium nanoparticles: A valuable catalyst for the synthesis of alkynyl ketones via acyl sonogashira reactions. *Catal. Lett.* **2020**, *150*, 652–659. [CrossRef]
54. Lukevics, E.; Abele, E.; Ignatovich, L. Biologically active silacyclanes. In *Advances in Heterocyclic Chemistry*; Katritzky, A.R., Ed.; Academic Press, Elsevier: Amsterdam, The Netherlands, 2010; Volume 99, pp. 107–141.
55. Ojima, I.; Vidal, E.S. Extremely regioselective intramolecular silylformylation of bis(silylamino)alkynes. *Organometallics* **1999**, *18*, 5103–5107. [CrossRef]
56. Ojima, I.; Vidal, E.; Tzamarioudaki, M.; Matsuda, I. Extremely regioselective intramolecular silylformylation of alkynes. *J. Am. Chem. Soc.* **1995**, *117*, 6797–6798. [CrossRef]
57. Aronica, L.A. Hydrosilylation and Silylformylation of Aceytlenic Compounds. Ph.D. Thesis, University of Pisa, Pisa, Italy, 1999.
58. Bonafoux, D.; Ojima, I. Desymmetrization of 4-dimethylsiloxy-1,6-heptadiynes through sequential double silylformylation. *Org. Lett.* **2001**, *3*, 1303–1305. [CrossRef]
59. Bonafoux, D.; Ojima, I. Novel DMAP-Catalyzed Skeletal Rearrangement of 5-exo-(2-Hydroxyethylene)oxasilacyclopentanes. *Org. Lett.* **2001**, *3*, 2333–2335. [CrossRef] [PubMed]
60. Denmark, S.E.; Kobayashi, T. Tandem Intramolecular silylformylation and silicon-assisted cross-coupling reactions. Synthesis of geometrically defined α,β-unsaturated aldehydes. *J. Org. Chem.* **2003**, *68*, 5153–5159. [CrossRef] [PubMed]
61. Zacuto, M.J.; O'Malle, S.J.; Leighton, J.L. Tandem intramolecular silylformylation–crotylsilylation: Highly efficient synthesis of polyketide fragments. *J. Am. Chem. Soc.* **2002**, *124*, 7890–7891. [CrossRef]
62. Zacuto, M.J.; O'Malley, S.J.; Leighton, J.L. Tandem silylformylation–allyl(crotyl)silylation: A new approach to polyketide synthesis. *Tetrahedron* **2003**, *59*, 8889–8900. [CrossRef]
63. Spletstoser, J.T.; Zacuto, M.J.; Leighton, J.L. Tandem silylformylation–crotylsilylation/tamao oxidation of internal alkynes: A remarkable example of generating complexity from simplicity. *Org. Lett.* **2008**, *10*, 5593–5596. [CrossRef] [PubMed]

64. Kim, H.; Ho, S.; Leighton, J.L. A More comprehensive and highly practical solution to enantioselective aldehyde crotylation. *Org. Lett.* **2008**, *10*, 5593–5596. [CrossRef] [PubMed]
65. Harrison, T.J.; Rabbat, P.M.A.; Leighton, J.L. An "aprotic" tamao oxidation/syn-selective tautomerization reaction for the efficient synthesis of the C(1)–C(9) fragment of fludelone. *Org. Lett.* **2012**, *14*, 4890–4893. [CrossRef]
66. Foley, C.N.; Leighton, J.L. Beyond the roche ester: A new approach to polypropionate stereotriad synthesis. *Org. Lett.* **2014**, *16*, 1180–1183. [CrossRef]
67. Foley, C.N.; Leighton, J.L. A Highly stereoselective, efficient, and scalable synthesis of the C(1)–C(9) Fragment of the epothilones. *Org. Lett.* **2015**, *17*, 5858–5861. [CrossRef]
68. Ojima, I. New cyclization reactions in organic syntheses. *Pure Appl. Chem.* **2002**, *74*, 159. [CrossRef]
69. Ojima, I.; Moralee, A.C.; Vassar, V.C. Recent advances in rhodium-catalyzed cyclization reactions. *Top. Catal.* **2002**, *19*, 89–99. [CrossRef]
70. Varchi, G.; Ojima, I. Synthesis of heterocycles through hydrosilylation, silylformylation, silylcarbocyclization and cyclohydrocarbonylation reactions. *Curr. Org. Chem.* **2006**, *10*, 1341–1362. [CrossRef]
71. Ojima, I.; Donovan, R.J.; Shay, W.R. Silylcarbocyclization reactions catalyzed by rhodium and rhodium-cobalt complexes. *J. Am. Chem. Soc.* **1992**, *114*, 6580–6582. [CrossRef]
72. Ojima, I.; McCullagh, J.V.; Shay, W.R. New cascade silylcarbocyclization (SiCaC) of enediynes. *J. Organomet. Chem.* **1996**, *521*, 421–423. [CrossRef]
73. Ojima, I.; Vu, A.T.; Lee, S.-Y.; McCullagh, J.V.; Moralee, A.C.; Fujiwara, M.; Hoang, T.H. Rhodium-catalyzed silylcarbocyclization (SiCaC) and carbonylative silylcarbocyclization (CO–SiCaC) reactions of enynes. *J. Am. Chem. Soc.* **2002**, *124*, 9164–9174. [CrossRef] [PubMed]
74. Fukuta, Y.; Matsuda, I.; Itoh, K. Rhodium-catalyzed domino silylformylation of enynes involving carbocyclization. *Tetrahedron Lett.* **1999**, *40*, 4703–4706. [CrossRef]
75. Park, K.H.; Jung, I.G.; Kim, S.Y.; Chung, Y.K. Immobilized cobalt/rhodium heterobimetallic nanoparticle-catalyzed silylcarbocyclization and carbonylative silylcarbocyclization of 1,6-enynes. *Org. Lett.* **2003**, *5*, 4967–4970. [CrossRef] [PubMed]
76. Murai, T.; Toshio, R.; Mutoh, Y. Sequential addition reaction of lithium acetylides and Grignard reagents to thioiminium salts from thiolactams leading to 2,2-disubstituted cyclic amines. *Tetrahedron* **2006**, *62*, 6312–6320. [CrossRef]
77. Denmark, S.E.; Liu, J.H.-C. Sequential silylcarbocyclization/silicon-based cross-coupling reactions. *J. Am. Chem. Soc.* **2007**, *129*, 3737–3744. [CrossRef] [PubMed]
78. Denmark, S.E.; Liu, J.H.-C.; Muhuhi, J.M. Total syntheses of isodomoic acids G and H. *J. Am. Chem. Soc.* **2009**, *131*, 14188–14189. [CrossRef] [PubMed]
79. Denmark, S.E.; Liu, J.H.-C.; Muhuhi, J.M. Stereocontrolled total syntheses of isodomoic acids G and H via a unified strategy. *J. Org. Chem.* **2011**, *76*, 201–215. [CrossRef]
80. Ojima, I.; Fracchiolla, D.A.; Donovan, R.J.; Banerji, P. Silylcarbobicyclization of 1,6-diynes: A novel Catalytic route to bicyclo[3.3.0]octenones. *J. Org. Chem.* **1994**, *59*, 7594–7595. [CrossRef]
81. Ojima, I.; Kass, D.F.; Zhu, J. New and efficient catalytic route to bicyclo[3.3.0]octa-1,5-dien-3-ones. *Organometallics* **1996**, *15*, 5191–5195. [CrossRef]
82. Ojima, I.; Zhu, J.; Vidal, E.S.; Kass, D.F. Silylcarbocyclizations of 1,6-diynes. *J. Am. Chem. Soc.* **1998**, *120*, 6690–6697. [CrossRef]
83. Matsuda, I.; Ishibashi, H.; Ii, N. Silylative cyclocarbonylation of acetylenic bonds catalyzed by $Rh_4(CO)_{12}$: An easy access to bicyclo[3.3.0]octenone skeletons. *Tetrahedron Lett.* **1995**, *36*, 241–244. [CrossRef]
84. Shibata, T.; Kadowaki, S.; Takagi, K. Chemo-and regioselective intramolecular hydrosilylative carbocyclization of allenynes. *Organometallics* **2004**, *23*, 4116–4120. [CrossRef]
85. Ojima, I.; Lee, S.-Y. Rhodium-catalyzed novel carbonylative carbotricyclization of enediynes. *J. Am. Chem. Soc.* **2000**, *122*, 2385–2386. [CrossRef]
86. Bennacer, B.; Fujiwara, M.; Ojima, I. Novel [2 + 2 + 2 + 1] Cycloaddition of enediynes catalyzed by rhodium complexes. *Org. Lett.* **2004**, *6*, 3589–3591. [CrossRef]
87. Bennacer, B.; Fujiwara, M.; Lee, S.-Y.; Ojima, I. Silicon-initiated carbonylative carbotricyclization and [2+2+2+1] cycloaddition of enediynes catalyzed by rhodium complexes. *J. Am. Chem. Soc.* **2005**, *127*, 17756–17767. [CrossRef]

88. Ojima, I.; Vu, A.T.; McCullagh, J.V.; Kinoshita, A. Rhodium-catalyzed intramolecular silylcarbotricyclization (SiCaT) of triynes. *J. Am. Chem. Soc.* **1999**, *121*, 3230–3231. [CrossRef]
89. Matsuda, I.; Ogiso, A.; Sato, S. Ready access of .alpha.-(triorganosilyl)methylene .beta.-lactones by means of rhodium-catalyzed cyclocarbonylation of substituted propargyl alcohols. *J. Am. Chem. Soc.* **1990**, *112*, 6120–6121. [CrossRef]
90. Matsuda, I.; Sakakibara, J.; Nagashima, H. A novel approach to α-silylmethylene-β-lactams via Rh-catalyzed silylcarbonylation of propargylamine derivatives. *Tetrahedron Lett.* **1991**, *32*, 7431–7434. [CrossRef]
91. Fukuta, Y.; Matsuda, I.; Itoh, K. Synthesis of 3-silyl-2(5H)-furanone by rhodium-catalyzed cyclocarbonylation. *Tetrahedron Lett.* **2001**, *42*, 1301–1304. [CrossRef]
92. Aronica, L.A.; Mazzoni, C.; Caporusso, A.M. Synthesis of functionalised β-lactones via silylcarbocyclisation/desilylation reactions of propargyl alcohols. *Tetrahedron* **2010**, *66*, 265–273. [CrossRef]
93. Aronica, L.A.; Caporusso, A.M.; Evangelisti, C.; Botavina, M.; Alberto, G.; Martra, G. Metal vapour derived supported rhodium nanoparticles in the synthesis of β-lactams and β-lactones derivatives. *J. Organomet. Chem.* **2012**, *700*, 20–28. [CrossRef]
94. Albano, G.; Morelli, M.; Aronica, L.A. Synthesis of functionalised 3-isochromanones by silylcarbocyclisation/desilylation reactions. *Eur. J. Org. Chem.* **2017**, *2017*, 3473–3480. [CrossRef]
95. Kong, K.-F.; Schneper, L.; Mathee, K. Beta-lactam antibiotics: From antibiosis to resistance and bacteriology. *APMIS* **2010**, *118*, 1–36. [CrossRef]
96. Shahid, M.; Sobia, F.; Singh, A.; Malik, A.; Khan, H.M.; Jonas, D.; Hawkey, P.M. Beta-lactams and Beta-lactamase-inhibitors in current-or potential-clinical practice: A comprehensive update. *Crit. Rev. Microbiol.* **2009**, *35*, 81–108. [CrossRef]
97. Romano, A.; Gaeta, F.; Poves, M.F.A.; Valluzzi, R.L. Cross-reactivity among beta-lactams. *Curr. Allergy Asthma Rep.* **2016**, *16*, 24. [CrossRef]
98. Broccolo, F.; Cainelli, G.; Caltabiano, G.; Cocuzza, C.E.A.; Fortuna, C.G.; Galletti, P.; Giacomini, D.; Musumarra, G.; Musumeci, R.; Quintavalla, A. Design, synthesis, and biological evaluation of 4-alkyliden-beta lactams: New products with promising antibiotic activity against resistant bacteria. *J. Med. Chem.* **2006**, *49*, 2804–2811. [CrossRef] [PubMed]
99. Bouthillier, G.; Mastalerz, H.; Menard, M.; Fung-Tomc, J.; Gradelski, E. The synthesis, antibacterial, and beta-lactamase inhibitory activity of a novel asparenomycin analog. *J. Antibiot.* **1992**, *45*, 240–245. [CrossRef] [PubMed]
100. Sinner, E.K.; Lichstrahl, M.S.; Li, R.; Marous, D.R.; Townsend, C.A. Methylations in complex carbapenem biosynthesis are catalyzed by a single cobalamin-dependent radical S-adenosylmethionine enzyme. *Chem. Commun.* **2019**, *55*, 14934–14937. [CrossRef] [PubMed]
101. Rolinson, G.N.; Geddes, A.M. The 50th anniversary of the discovery of 6-aminopenicillanic acid (6-APA). *Int. J. Antimicrob. Agents* **2007**, *29*, 3–8. [CrossRef] [PubMed]
102. Golemi, D.; Maveyraud, L.; Ishiwata, A.; Tranier, S.; Miyashita, K.; Nagase, T.; Massova, I.; Mourey, L.; Samama, J.P.; Mobashery, S. 6-(hydroxyalkyl)penicillanates as probes for mechanisms of beta-lactamases. *J. Antibiot.* **2000**, *53*, 1022–1027. [CrossRef]
103. Liang, Y.; Raju, R.; Le, T.; Taylor, C.D.; Howell, A.R. Cross-metathesis of α-methylene-β-lactams: The first tetrasubstituted alkenes by CM. *Tetrahedron Lett.* **2009**, *50*, 1020–1022. [CrossRef]
104. Zhu, L.; Xiong, Y.; Li, C. Synthesis of α-methylene-β-lactams via PPh3-catalyzed umpolung cyclization of propiolamides. *J. Org. Chem.* **2015**, *80*, 628–633. [CrossRef]
105. Hussein, M.; Nasr El Dine, A.; Farès, F.; Dorcet, V.; Hachem, A.; Grée, R. A new direct synthesis of α-methylene- and α-alkylidene-β-lactams. *Tetrahedron Lett.* **2016**, *57*, 1990–1993. [CrossRef]
106. Zhang, L.; Ma, L.; Zhou, H.; Yao, J.; Li, X.; Qiu, G. Synthesis of α-methylene-β-lactams enabled by base-promoted intramolecular 1,2-addition of N-propiolamide and C–C bond migrating cleavage of aziridine. *Org. Lett.* **2018**, *20*, 2407–2411. [CrossRef]
107. Ojima, I.; Machnik, D.; Donovan, R.J.; Mneimne, O. Silylcyclocarbonylation (SiCCa) of alkynylamines catalyzed by rhodium complexes. *Inorg. Chim. Acta* **1996**, *251*, 299–307. [CrossRef]
108. Ojima, I.; Clos, N.; Donovan, R.J.; Ingallina, P. Hydrosilylation of 1-hexyne catalyzed by rhodium and cobalt-rhodium mixed-metal complexes. Mechanism of apparent trans addition. *Organometallics* **1990**, *9*, 3127–3133. [CrossRef]

109. Albano, G.; Morelli, M.; Lissia, M.; Aronica, L.A. Synthesis of functionalised indoline and isoquinoline derivatives through a silylcarbocyclisation/ desilylation sequence. *ChemistrySelect* **2019**, *4*, 2505–2511. [CrossRef]
110. Xu, X.; Doyle, M.P. The [3 + 3]-cycloaddition alternative for heterocycle syntheses: Catalytically generated metalloenolcarbenes as dipolar adducts. *Acc. Chem. Res.* **2014**, *47*, 1396–1405. [CrossRef] [PubMed]
111. Albano, G.; Aronica, L.A. Cyclization reactions for the synthesis of phthalans and isoindolines. *Synthesis* **2018**, *50*, 1209–1227.
112. Pucci, A.; Albano, G.; Pollastrini, M.; Lucci, A.; Colalillo, M.; Oliva, F.; Evangelisti, C.; Marelli, M.; Santalucia, D.; Mandoli, A. Supported tris-triazole ligands for batch and continuous-flow copper-catalyzed huisgen 1,3-dipolar cycloaddition reactions. *Catalysts* **2020**, *10*, 434. [CrossRef]
113. Albano, G.; Aronica, L.A. Potentiality and synthesis of o-and n-heterocycles: Pd-catalyzed cyclocarbonylative sonogashira coupling as a valuable route to phthalans, isochromans, and isoindolines. *Eur. J. Org. Chem.* **2017**, *2017*, 7204–7221. [CrossRef]
114. Aronica, L.A.; Albano, G.; Giannotti, L.; Meucci, E. Synthesis of n-heteroaromatic compounds through cyclocarbonylative sonogashira reactions. *Eur. J. Org. Chem.* **2017**, *2017*, 955–963. [CrossRef]
115. Albano, G.; Aronica, L.A. Acyl sonogashira cross-coupling: State of the art and application to the synthesis of heterocyclic compounds. *Catalysts* **2020**, *10*, 25. [CrossRef]
116. Albano, G.; Giuntini, S.; Aronica, L.A. Synthesis of 3-Alkylideneisoindolin-1-ones via sonogashira cyclocarbonylative reactions of 2-ethynylbenzamides. *J. Org. Chem.* **2020**, *85*, 10022–10034. [CrossRef]
117. Konishi, H.; Manabe, K. Formic acid derivatives as practical carbon monoxide surrogates for metal-catalyzed carbonylation reactions. *Synlett* **2014**, *25*, 1971–1986. [CrossRef]
118. Wu, L.; Liu, Q.; Jackstell, R.; Beller, M. Carbonylations of alkenes with CO surrogates. *Angew. Chem. Int. Ed.* **2014**, *53*, 6310–6320. [CrossRef] [PubMed]
119. Gautam, P.; Bhanage, B.M. Recent advances in the transition metal catalyzed carbonylation of alkynes, arenes and aryl halides using CO surrogates. *Catal. Sci. Technol.* **2015**, *5*, 4663–4702. [CrossRef]

© 2020 by the authors. Licensee MDPI, Basel, Switzerland. This article is an open access article distributed under the terms and conditions of the Creative Commons Attribution (CC BY) license (http://creativecommons.org/licenses/by/4.0/).

MDPI
St. Alban-Anlage 66
4052 Basel
Switzerland
Tel. +41 61 683 77 34
Fax +41 61 302 89 18
www.mdpi.com

Catalysts Editorial Office
E-mail: catalysts@mdpi.com
www.mdpi.com/journal/catalysts

www.ingramcontent.com/pod-product-compliance
Lightning Source LLC
LaVergne TN
LVHW070552100526
838202LV00012B/447